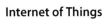

Internet of Things

Internet of Things

Architectures, Protocols and Standards

Simone Cirani
Caligoo Inc., Chicago, IL, USA

Gianluigi Ferrari
Department of Engineering and Architecture
University of Parma, Parma (PR), Italy

Marco Picone
Caligoo Inc., Chicago, IL, USA

Luca Veltri
Department of Engineering and Architecture
University of Parma, Parma (PR), Italy

Registered Offices
John Wiley & Sons, Inc., 111 River Street, Hoboken, NJ 07030, USA
John Wiley & Sons Ltd, The Atrium, Southern Gate, Chichester, West Sussex, PO19 8SQ, UK

Editorial Office
The Atrium, Southern Gate, Chichester, West Sussex, PO19 8SQ, UK

For details of our global editorial offices, customer services, and more information about Wiley products visit us at www.wiley.com.

Wiley also publishes its books in a variety of electronic formats and by print-on-demand. Some content that appears in standard print versions of this book may not be available in other formats.

Library of Congress Cataloging-in-Publication Data

Names: Cirani, Simone, 1982- author. | Ferrari, Gianluigi, 1974- author. | Picone, Marco, 1984- author. | Veltri, Luca, author.
Title: Internet of things : architectures, protocols and standards / Simone Cirani, Taneto, Gattatico (RE), Italy, Ph.D., Gianluigi Ferrari, Ph.D., Marco Picone, Gattatico (RE), Italy Ph.D., Luca Veltri, Ph.D., Parma (PR), Italy.
Description: First edition. | Hoboken, NJ : Wiley, 2019. |
Identifiers: LCCN 2018021870 (print) | LCCN 2018028978 (ebook) | ISBN 9781119359685 (Adobe PDF) | ISBN 9781119359708 (ePub) | ISBN 9781119359678 (hardcover)
Subjects: LCSH: Internet of things.
Classification: LCC TK5105.8857 (ebook) | LCC TK5105.8857 .C55 2019 (print) | DDC 004.67/8–dc23
LC record available at https://lccn.loc.gov/2018021870

Cover design by Wiley
Cover image: © shulz/iStockphoto

Set in 10/12pt WarnockPro by SPi Global, Chennai, India

10 9 8 7 6 5 4 3 2 1

"Machines take me by surprise with great frequency."
(Alan Mathison Turing)

I would like to dedicate this work: To Paola, the love of my life, my hero, my strength. You are what I live for. To my wonderful mom and dad. You have always supported me. You gave me everything. You taught me the value of work and commitment. You gifted me with your love. You are always in my heart and thoughts. To my fantastic sisters, who have always been an example. You believed in me and blessed me with your love, joy, and laughs. To my grandma, you will always have a special place in my heart. You are an incredible inspiration. I miss you. To Jonas, who has taught me the passion for knowledge, exploration, and science. To Emil and Emma, I wish you all the best. I am proud of you. I will never say thanks enough. I love you all. Thanks to Marco, Gianluigi, and Luca, I am really proud and honored to have worked and researched with you. I am so proud of what we have achieved in these years. Finally, thanks to all my colleagues, old and new, who contributed to make this book happen in one way or another.
Simone Cirani

To the women of my life, Anna, Sofia, and Viola: You fill my heart and brighten my days.
Gianluigi Ferrari

To Eleonora and my parents, Antonio and Marina, who are always by my side in every choice and decision. A special thanks to all the people who worked with us, supported our vision, and shared the challenges during these years.
Marco Picone

To my family.
Luca Veltri

Contents

Preface

The Internet of Things or, as commonly referred to and now universally used, IoT has two keywords: things and Internet. The very idea of IoT consists allowing things to connect to the (existing) Internet, thus allowing the generation of information and, on the reverse, the interaction of the virtual world with the physical world. This book does not attempt to be an exhaustive treaty on the subject of IoT. Rather, it tries to present a broad view of the IoT based on the joint research activity at the University of Parma, mainly in the years between 2012 and 2015 (when all the authors were affiliated with the same Department of Information Engineering), especially in the context of the EU FP7 project CALIPSO (Connect All IP-based Smart Objects!, 2012–2014). In particular, we present, in a coherent way, new ideas we had the opportunity to explore in the IoT ecosystem, trying to encompass the presence of heterogeneous communication technologies through unifying concepts such as interoperability, discoverability, security, and privacy. On the way, we also touch upon cloud and fog computing (two concepts interwoven with IoT) and conclude with a practical view on IoT (with focus on the physical devices). The intended audience of the book is academic and industrial professionals, with good technical skills in networking technologies. To ease reading, we have tried to provide intuition behind all presented concepts.

The contents of the book flow from a preliminary overview on the Internet and the IoT, with details on "classical" protocols, to more technical details. The synopsis of the book can be summarized as follows: The *first chapter* introduces IoT in general terms and illustrates a few IoT-enabled applications, from home/building automation to smart farming. The *second chapter* contains an overview of relevant standards (e.g. Constrained Application Protocol, CoAP), presented according to the protocol layers and parallelizing the "traditional" Internet and the IoT, with a final outlook on industrial IoT. *Chapter three* focuses on interoperability, a key concept for IoT, highlighting relevant aspects (e.g. Representational State Transfer (REST) architectures and Web of Things) and presenting illustrative applications (e.g. the Dual-network Management Protocol (DNMP) allowing the interaction of IEEE 802.11s and IEEE 802.15.4 networks). At the end of Chapter three, we preliminarily also discuss discoverability in constrained environments (with reference to the CoRE Link Format); this paves the way to *Chapter four*, which dives into the concept of discoverability (both in terms of service and resource discovery), presenting a few of our research results in this area. *Chapter five* is dedicated to security and privacy in the IoT, discussing proper mechanisms for IoT in a comparative way with respect to common mechanisms for classical Internet. In *Chapter six*, we consider cloud and fog computing, discussing concepts such as big stream processing (relevant for cloud-based applications) and the IoT Hub (relevant for fog-based applications). Finally, *Chapter seven* is an overview of hands-on issues, presenting relevant hardware devices and discussing a Web-of-Things-oriented vision for a test bed implementation.

We remark that the specific IoT protocols, algorithms, and architectures considered in this book are "representative," as opposed to "universal." In other words, we set to write this book mainly to provide the reader with our vision on IoT. Our hope is that this book will be interpreted as a starting point and a useful comparative reference for those interested in the continuously evolving subject of the IoT.

It is our pleasure to thank all the collaborators and students who were with us during the years of research that have led to this book, collaborating with the Wireless Adhoc and Sensor Networks (WASN) Lab of Department of Information Engineering of the University of Parma, which has lately been "rebranded," owing to this intense research activity, as the IoT Lab at the Department of Engineering and Architecture. We particularly thank, for fundamental contributions, Dr. Laura Belli, Dr. Luca Davoli, Dr. Paolo Medagliani, Dr. Stefano Busanelli, Gabriele Ferrari, Vincent Gay, Dr. Jérémie Leguay, Mattia Antonini, Dr. Andrea Gorrieri, Lorenzo Melegari, and Mirko Mancin. We also thank, for collaborative efforts and useful discussions, Dr. Michele Amoretti, Dr. Francesco Zanichelli, Dr. Andrzej Duda, Dr. Simon Duquennoy, Dr. Nicola Iotti, Dr. Andrea G. Forte, and Giovanni Guerri. Finally, we express our sincere gratitude to Wiley for giving us the opportunity to complete this project. In particular, we are indebted to Tiina Wigley, our executive commissioning editor, for showing initial interest in our proposal; we are *really* indebted to Sandra Grayson, our associate book editor, who has shown remarkable patience and kindness, tolerating our delay and idiosyncrasies throughout the years of writing.

Parma, July 2018

Simone Cirani
Gianluigi Ferrari
Marco Picone
Luca Veltri

1

Preliminaries, Motivation, and Related Work

1.1 What is the Internet of Things?

The Internet of Things (IoT) encapsulates a vision of a world in which billions of objects with embedded intelligence, communication means, and sensing and actuation capabilities will connect over IP (Internet Protocol) networks. Our current Internet has undergone a fundamental transition, from hardware-driven (computers, fibers, and Ethernet cables) to market-driven (Facebook, Amazon) opportunities. This has come about due to the interconnection of seamingly disjoint intranets with strong horizontal software capabilities. The IoT calls for open environments and an integrated architecture of interoperable platforms. Smart objects and cyber-physical systems – or just "things" – are the new IoT entities: the objects of everyday life, augmented with micro-controllers, optical and/or radio transceivers, sensors, actuators, and protocol stacks suitable for communication in constrained environments where target hardware has limited resources, allowing them to gather data from the environment and act upon it, and giving them an interface to the physical world. These objects can be worn by users or deployed in the environment. They are usually highly constrained, with limited memory and available energy stores, and they are subject to stringent low-cost requirements. Data storage, processing, and analytics are fundamental requirements, necessary to enrich the raw IoT data and transform them into useful information. According to the "Edge Computing" paradigm, introducing computing resources at the edge of access networks may bring several benefits that are key for IoT scenarios: low latency, real-time capabilities and context-awareness. Edge nodes (servers or micro data-centers on the edge) may act as an interface to data

Internet of Things: Architectures, Protocols and Standards, First Edition.
Simone Cirani, Gianluigi Ferrari, Marco Picone, and Luca Veltri.

streams coming from connected devices, objects, and applications. The stored Big Data can then be processed with new mechanisms, such as machine and deep learning, transforming raw data generated by connected objects into useful information. The useful information will then be disseminated to relevant devices and interested users or stored for further processing and access.

1.2 Wireless Ad-hoc and Sensor Networks: The Ancestors without IP

Wireless sensor networks (WSNs) were an emerging application field of microelectronics and communications in the first decade of the twenty-first century. In particular, WSNs promised wide support of interactions between people and their surroundings. The potential of a WSN can be seen in the three words behind the acronym:

- "Wireless" puts the focus on the freedom that the elimination of wires gives, in terms of mobility support and ease of system deployment;
- "Sensor" reflects the capability of sensing technology to provide the means to perceive and interact — in a wide sense — with the world;
- "Networks" gives emphasis to the possibility of building systems whose functional capabilities are given by a plurality of communicating devices, possibly distributed over large areas.

Pushed on by early military research, WSNs were different from traditional networks in terms of the communication paradigm: the address-centric approach used in end-to-end transmissions between specific devices, with explicit indication of both source and destination addresses in each packet, was to be replaced with an alternative (and somewhat new) data-centric approach. This "address blindness" led to the selection of a suitable data diffusion strategy – in other words, communication protocol – for data-centric networks. The typical network deployment would consist of the sources placed around the areas to be monitored and the sinks located in easily accessible places. The sinks provided adequate storage capacity to hold the data from the sources. Sources might send information to sinks in accordance with different scheduling policies: periodic (i.e., time-driven), event specific (i.e., event-driven), a reply in response to requests coming from sinks (i.e., query-driven), or some combination thereof.

Because research focused on the area, WSNs have typically been associated with ad-hoc networks, to the point that the two terms have almost become – although erroneously so – synonymous. In particular, ad-hoc networks are defined as general, infrastructure-less, cooperation-based, opportunistic networks, typically customized for specific scenarios and applications. These kinds of networks have to face frequent and random variations of many factors (radio channel, topology, data traffic, and so on), implying a need for dynamic management of a large number of parameters in the most efficient, effective, and reactive way. To this end, a number of key research problems have been studied, and solutions proposed, in the literature:

- self-configuration and self-organization in infrastructure-less systems;
- support for cooperative operations in systems with heterogenous members;
- multi-hop peer-to-peer communication among network nodes, with effective routing protocols;
- network self-healing behavior providing a sufficient degree of robustness and reliability;
- seamless mobility management and support of dynamic network topologies.

1.3 IoT-enabled Applications

The IoT touches every facet of our lives. IoT-enabled applications are found in a large number of scenarios, including: home and building automation, smart cities, smart grids, Industry 4.0, and smart agriculture. In each of these areas, the use of a common (IP-oriented) communication protocol stack allows the building of innovative applications. In this section, we provide a concise overview of potential applications in each of these areas.

1.3.1 Home and Building Automation

As the smart home market has seen growing investment and has continued to mature, ever more home automation applications have appeared, each designed for a specific audience. The result has been the creation of several disconnected vertical market segments. Typical examples of increasingly mainstream applications are related to home security and energy efficiency and energy saving. Pushed by the innovations in light and room control, the IoT will foster the development of endless applications for home automation. For example, a typical example of an area of home automation that is destined to grow in the context of the IoT is in healthcare, namely IoT-enabled solutions for the physically less mobile (among others, the elderly, particulary relevant against a background of aging populations), and for the disabled or chronically ill (for instance, remote health monitoring and air-quality monitoring). In general, building automation solutions are starting to converge and are also moving, from the current applications in luxury, security and comfort, to a wider range of applications and connected solutions; this will create market opportunities. While today's smart home solutions are fragmented, the IoT is expected to lead to a new level of interoperability between commercial home and building automation solutions.

1.3.2 Smart Cities

Cities are complex ecosystems, where quality of life is an important concern. In such urban environments, people, companies and public authorities experience specific needs and demands in domains such as healthcare, media, energy and the environment, safety, and public services. A city is perceived more and more as being like a single "organism", which needs to be efficiently monitored to provide citizens with accurate information. IoT technologies are fundamental to collecting data on the city status and disseminating them to citizens. In this context, cities and urban areas represent a critical mass when it comes to shaping the demand for advanced IoT-based services.

1.3.3 Smart Grids

A smart grid is an electrical grid that includes a variety of operational systems, including smart meters, smart appliances, renewable energy resources, and energy-efficient resources. Power line communications (PLC) relate to the use of existing electrical cables to transport data and have been investigated for a long time. Power utilities have been using this technology for many years to send or receive (limited amounts of) data on the existing power grid. Although PLC is mostly limited by the type of propagation medium, it can use existing wiring in the distribution network. According to EU's standards and laws, electrical utility companies can use PLC for low bit-rate data transfers (with data rates lower than 50 Kbps) in the 3–148 kHz frequency band. This technology opens up new opportunities and new forms of interactions among people and things in many application areas, such as smart metering services and energy consumption reporting. This makes PLC an enabler for sensing, control, and automation in large systems spread over relatively wide areas, such as in the smart city and smart grid scenarios. On top of PLC, one can also adopt enabling technologies that can improve smart automation processes, such as the IoT. For instance, the adoption of the PLC technology in industrial scenarios (e.g., remote control in automation and manufacturing companies), paves the way to the "Industrial IoT". Several applications have been enabled by PLC technology's ability to recover from network changes (in terms of repairs and improvements, physical removal, and transfer function) mitigating the fallout on the signal transmission.

Nevertheless, it is well known that power lines are far from ideal channels for data transmission (due to inner variations in location, time, frequency band and type of equipment connected to the line). As a result there has been increasing interest in the joint adoption of IoT and PLC paradigms to improve the robustness of communication. This has led to the suggestion of using small, resource-constrained devices (namely, IoT), with pervasive computing capabilities, and internet standard solutions (as proposed by Internet standardization organizations, such as IETF, ETSI and W3C). Such systems can be key components for implementing future smart grids.

1.3.4 Industrial IoT

The Industrial Internet of Things (IIoT) describes the IoT as used in industries such as manufacturing, logistics, oil and gas, transportation, energy/utilities, mining and metals, aviation and others. These industries represent the majority of gross domestic product among the G20 nations. The IIoT is still at an early stage, similar to where the Internet was in the late 1990s. While the evolution of the consumer Internet over the last two decades provides some important lessons, it is unclear how much of this learning is applicable to the IIoT, given its unique scope and requirements. For example, *real-time* responses are often critical in manufacturing, energy, transportation and healthcare: real time for today's Internet usually means a few seconds, whereas real time for industrial machines involves sub-millisecond scales. Another important consideration is reliability. The current Internet embodies a "best effort" approach, which provides acceptable performance for e-commerce or human interactions. However, the failure of the power grid, the air traffic control system, or an automated factory for the same length of time would have much more serious consequences.

Much attention has been given to the efforts of large companies such as Cisco, GE, and Huawei, and government initiatives such as Industrie 4.0 in Germany. For example:

- GE announced that it realized more than $1 billion in incremental revenues in 2014 by helping customers improve asset performance and business operations through IIoT capabilities and services.
- The German government is sponsoring "Industrie 4.0", a multi-year strategic initiative that brings together leaders from the public and private sectors as well as from academia to create a comprehensive vision and action plan for applying digital technologies to the German industrial sector.
- Other European countries have their own industrial transformation projects in which the IIoT takes center stage, such as Smart Factory (the Netherlands), Industry 4.0 (Italy), Industry of the Future (France), and others.
- China has also recently launched its "Made in China 2025" strategy to promote domestic integration of digital technologies and industrialization.

As the IIoT gains momentum, one of the biggest bottlenecks faced is the inability to share information between smart devices that may be speaking different "languages". This communication gap stems from the multiple protocols used on factory floors. So, while you can put a sensor on a machine to gather data, the ability to push that information across a network and ultimately "talk" with other systems is a bit more difficult. Standardization is therefore a key aspect of the IIoT.

The IIoT's potential payoff is enormous. Operational efficiency is one of its key attractions, and early adopters are focused on these benefits. By introducing automation and more flexible production techniques, for instance, manufacturers could boost their productivity by as much as 30%. In this context, three IIoT capabilities must be mastered:

- *sensor-driven computing*: converting sensed data into insights (using the industrial analytics described below) that operators and systems can act on;
- *industrial analytics*: turning data from sensors and other sources into actionable insights;
- *intelligent machine applications*: integrating sensing devices and intelligent components into machines.

1.3.5 Smart Farming

Modern agriculture is facing tremendous challenges as it attempts to build a sustainable future across different regions of the globe. Examples of such challenges include population increase, urbanization, an increasingly degraded environment, an increasing trend towards consumption of animal proteins, changes in food preferences as a result of aging populations and migration, and of course climate change. A modern agriculture needs to be developed, characterized by the adoption of production processes, technologies and tools derived from scientific advances, and results from research and development activities.

Precision farming or smart agriculture is an area with the greatest opportunities for digital development but with the lowest penetration, to date, of digitized solutions. The farming industry will become arguably more important than ever before in the next few decades. It could derive huge benefits from the use of environmental and terrestrial sensors, applications for monitoring the weather, automation for more precise application of fertilizers and pesticides (thus reducing waste of natural resources), and the adoption of planning strategies for maintenance.

Smart farming is already becoming common, thanks to the application of new technologies, such as drones and sensor networks (to collect data) and cloud platforms (to manage the collected data). The set of technologies used in smart farming are as complex as the activities run by farmers, growers, and other stakeholders in the sector. There are is a wide spectrum of possible applications: fleet management, livestock monitoring, fish farming, forest care, indoor city farming, and many more. All of the technologies involved revolve around the concept of the IoT and aim at supporting farmers in their decision processes through decision-support systems. They involve real-time data at a level of granularity not previously possible. This enables better decisions to be made, translating into less waste and an increase in efficiency.

Communication technologies are a key component of smart agriculture applications. In particular, wireless communication technologies are attractive, because of the significant reduction and simplification in wiring involved. Various wireless standards have been established. One can group these into two main categories, depending on the transmission range:

- *Short-range communication*: including standards for:
  - wireless LAN, used for Wi-Fi, namely IEEE 802.11
  - wireless PAN, used more widely for measurement and automation applications, such as IEEE 802.15.1 (Bluetooth) (IEEE, 2002) and IEEE 802.15.4 (ZigBee/6LoWPAN) (IEEE, 2003).
  All these standards use the instrumentation, scientific and medical (ISM) radio bands, typically operating in the 2.400–2.4835 GHz band.
- *Long-range communication*: including the increasingly important sub-gigahertz IoT communication techologies, such as LoRA, in the 868–870 MHz band. These trade data transmission rates (on the order of hundreds of kbit/s) for longer transmission ranges.

Communication technologies can be also classified according to the specific application:

- environmental monitoring (weather monitoring and geo-referenced environmental monitoring)
- precision agriculture
- machine and process control (M2M communications)
- facility automation
- traceability systems.

2

Standards

2.1 "Traditional" Internet Review

The original idea of the Internet was that of connecting multiple independent networks of rather arbitrary design. It began with the ARPANET as the pioneering packet switching network, but soon included packet satellite networks, ground-based packet radio networks and other networks. The current Internet is based on the concept of open-architecture networking (an excellent overview of the history of the Internet is in an article by Leiner *et al.* [1]). According to this original approach, the choice of any individual network technology was not dictated by a particular network architecture but rather could be selected freely by a provider and made to interwork with the other networks through a meta-level "internetworking architecture". The use of the open systems interconnect (OSI) approach, with the use of a layer architecture, was instrumental in the design of interactions between different networks. The TCP/IP protocol suite has proven to be a phenomenally flexible and scalable networking strategy. Internet Protocol (IP) (layer three) provides only for addressing and forwarding of individual packets, while the transport control protocol (TCP; layer four), is concerned with service features such as flow control and recovery when there are lost packets. For those applications that do not need the services of TCP, the User Datagram Protocol (UDP) provides direct access to the basic service of IP.

In practice, the seven-layer architecture foreseen by the ISO-OSI protocol stack has been replaced by a five-layer IP stack. This is typically referred to as the TCP/IP protocol stack, because the TCP is the most-used protocol in the transport layer and IP is the almost ubiquitous in the network layer. The three upper layers of the

Internet of Things: Architectures, Protocols and Standards, First Edition.
Simone Cirani, Gianluigi Ferrari, Marco Picone, and Luca Veltri.

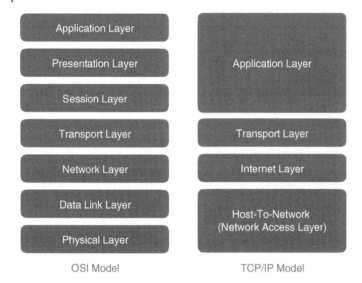

Figure 2.1 Communication protocol stacks: traditional seven-layer ISO-OSI stack (left) versus four-layer TCP/IP stack (right).

ISO-OSI protocol stack – the session (layer five), presentation (layer six), and application (layer seven) – converge in a single (fifth) layer in the TCP/IP protocol stack, namely The layered architecture of the Internet (according to the ISO-OSI and TCP/IP models) is shown in Figure 2.1.

In the following, we summarize the main communication protocols used in the various layers of the ISO-OSI communication protocol stack. In particular, we will outline:

- at the physical/link layer (L1/L2), the IEEE802.3 (Ethernet) and IEEE 802.11 (Wi-Fi) protocols;
- at the network layer (L3), IPv4 and IPv6;
- at the transport layer (L4), TCP and UDP;
- at the application layer (L5), Hypertext Transfer Protocol (HTTP) and Session Initiation Protocol (SIP).

2.1.1 Physical/Link Layer

In this subsection, we focus on two relevant communication protocols for physical/link (PHY/MAC) layers, namely the IEEE 802.3 standard (typically referred to as Ethernet) and the IEEE 802.11 (which refers to the vast family of Wi-Fi standards, with all their amendments). While the former applies to wired local area networks (LANs), the latter applies to wireless LANs (WLANs). Being related to the bottom two layers of the protocol stack, they mostly refer to point-to-point communications; in other words, there is no concept of routing.

2.1.1.1   IEEE 802.3 (Ethernet)

IEEE 802.3 is the set of standards issued by the Institute of Electrical and Electronics Engineers (IEEE) to define Ethernet-based networks as well as the name of the working group assigned to develop them. IEEE 802.3 is otherwise known as the Ethernet standard and defines the physical layer and the media access control (MAC) for the data link layer for wired Ethernet networks. It is generally a local area network (LAN) technology.

IEEE 802.3 specifies the physical and networking characteristics of an Ethernet network, such as how physical connections between nodes (routers/switches/hubs) are made through various wired media, such as copper coaxial or fiber cables. The technology was developed to work with the IEEE 802.1 standard for network architecture and the first released standard was Ethernet II in 1982, which featured 10 Mbit/s delivered over thick coaxial cable and frames with a type field. In 1983, the first standard with the name IEEE 802.3 was developed for 10BASE5 (also known as "thick Ethernet" or thicknet). It had the same speed as the earlier Ethernet II standard, but the type field was replaced by a length field. IEEE 802.3a followed in 1985, and was designated as 10BASE2, which was essentially the same as 10BASE5 but ran on thinner coaxial cables, therefore it was also known as "thinnet" or "cheapnet".

There are a multitude of additions and revisions to the 802.3 standard and each is designated by letters appended after the number "3". Other notable standards are 802.3i for 10Base-T for twister pair wire and 802.3j 10BASE-F for fiber-optic cables, with the latest revision (2016) being 802.3bz, which supports 2.5GBASE-T and 5GBASE-T: 2.5-Gbit and 5-Gbit Ethernet over Cat-5/Cat-6 twisted pair wires.

At layer two, Ethernet relies on carrier-sense multiple access with collision detection technology (CSMA/CD). A CSMA protocol works as follows. A station desiring to transmit senses the medium. If the medium is busy (i.e., some other station is transmitting) then the station defers its transmission to a later time. If the medium is sensed as being free then the station is allowed to transmit. CSMA is very effective when the medium is not heavily loaded since it allows stations to transmit with minimum delay. But there is always a chance of stations simultaneously sensing the medium as being free and transmitting at the same time, causing a collision. These collision situations must be identified so that the MAC layer can retransmit the frame by itself and not rely on the upper layers, which would cause significant delay. Ethernet relies on the CD mechanism to mitigate this condition. It uses a carrier-sensing scheme in which a transmitting station can detect collisions while transmitting a frame. It does this by sensing transmissions from other stations. When a collision condition is detected, the station stops transmitting that frame, transmits a jam signal, and then waits for a random time interval before trying to resend the frame. This collision detection approach is possible over cabled networks, but does not work in wireless networks. CSMA/CD improves the CSMA performance by terminating transmission as soon as a collision is detected, thus shortening the time required before a retry can be attempted.

2.1.1.2 IEEE 802.11

IEEE 802.11 is a set of PHY/MAC specifications for implementing wireless local area networks (WLAN) in various frequency bands, including the 900 MHz and the 2.4, 3.6, 5, and 60 GHz bands. The base version of the standard was released in 1997, and has had numerous subsequent amendments. The standard and its amendments provide the basis for wireless network products using the Wi-Fi brand. While each amendment is officially revoked when it is incorporated in the latest version of the standard, the corporate world tends to market the revisions individually, because they concisely denote the capabilities of their products. As a result, in the marketplace, each revision tends to become its own standard. Among the latest amendments are:

- *IEEE 802.11ac (2013)*: which guarantees very high throughput in the frequency band below 6 GHz, and brings potential improvements over 802.11n, including a better modulation scheme, wider channels, and multi-user MIMO;
- *IEEE 802.11ah (2016)*: for sub-GHz license-exempt operations, such as sensor networks and smart metering;
- *IEEE 802.11ai*: which introduces fast initial link setup.

An 802.11 LAN is based on a "cellular" architecture: the system is subdivided into cells. Each cell, referred to as a basic service set in the

802.11 nomenclature, is controlled by a base station, known as an access point (AP). Although a wireless LAN may be formed by a single cell, with a single AP, most installations are formed by several cells, with the APs connected through some backbone, denoted as the distribution system (DS). This backbone is typically an Ethernet, and in some cases is wireless itself. The whole interconnected WLAN, including the different cells, their respective APs and the DS, is seen as a single 802 network to the upper layers of the OSI model and is known as an extended service set.

The basic access mechanism, called the distributed coordination function, is basically a carrier sense multiple access with collision avoidance technology (CSMA/CA). As notes above, CD mechanisms are a good idea in a wired LAN, but they cannot be used in a WLAN environment for two main reasons:

- it would require the implementation of a full-duplex radio, increasing the price significantly;
- in a wireless environment we cannot assume that all stations hear each other (which is the basic assumption of the CD scheme), and the fact that a station wants to transmit and senses the medium as free does not necessarily mean that the medium *is* free around the receiver area.

In order to overcome these problems, the 802.11 standard uses a CA mechanism together with a positive acknowledgement scheme, as follows:

1) A station wanting to transmit senses the medium: if the medium is busy then it defers; if the medium is free for a specified time (referred to as the distributed interframe space), then the station is allowed to transmit.
2) The receiving station checks the cyclic redundancy check (CRC) of the received packet and sends an acknowledgment packet (ACK). Receipt of the ACK indicates to the transmitter that no collision occurred. If the sender does not receive the ACK then it retransmits the fragment until it receives the ACK or, if after a given number of retransmissions, no ACK is received, the packet is discarded.

Virtual carrier sensing is another mechanism used to reduce the probability of collisions between two stations that are not within transmission range of each other. A station wanting to transmit a packet first transmits a short control packet, referred to as a request to send (RTS). This includes the source, destination, and the duration of the following transaction; in other words, the packet and the respective ACK packet. The destination station then responds (if the medium is free) with a response control packet, referred to as the clear to send (CTS), which includes the same duration information. All stations receiving either the RTS and/or the CTS, set their virtual carrier sense indicators (referred to as the network allocation vector, NAV), for the given duration, and use this information together with the physical carrier sense when sensing the medium. This mechanism reduces the probability of a collision in the receiver area by a station that is "hidden" from the transmitter to the short duration of the RTS transmission. This is because the station hears the CTS and reserves the medium as busy until the end of the transaction. The duration information on the RTS also protects the transmitter area from collisions during the ACK (from stations that are out of range of the acknowledging station). It should also be noted that, because the RTS and CTS are short frames, the mechanism also reduces the overhead of collisions, since these are recognized faster than if the whole packet were to be transmitted – this is true if the packet is significantly bigger than the RTS, so the standard allows for short packets to be transmitted without the RTS/CTS transaction.

2.1.2 Network Layer

In this subsection, we focus on the key protocol at the network layer (layer 3), namely IP, the key protocol for relaying datagrams across the Internet, defined as a combination of heterogeneous networks. IP is thus the key protocol to enable inter-networking and to allow efficient and robust routing in a very scalable way. The current version of IP is version 4 (IPv4), which relies on 32-bit addresses. However, its designated successor and thr fundamental enabler of the IoT is IPv6, which used 128-bit addresses, thus allowing the number of addressable "things" to explode. In the following, a comparative oveview of IPv4 and IPv6 is presented.

2.1.2.1 IPv6 and IPv4
IPv6 is the next-generation Internet protocol, and the Internet is still in its transition from IPv4. IPv4 public addresses have been exhausted and various techniques – such as Dynamic Host Control Protocol (DHCP), network address translation (NAT), and sub-netting – have been proposed in order to slow down the rate at which IPv4 IP address exhaustion is approaching.

In practice, IPv6 is much more than an extension of IPv4 addressing. IPv6, first defined in the RFC 2460 standard, is a complete implementation of the network layer of the TCP/IP protocol stack and it covers a lot more than a simple address-space extension from 32 to 128 bits (the mechanism that gives IPv6 its ability to allocate addresses to all the devices in the world for decades to come).

The technical functioning of the Internet remains the same with both versions of IP, and it is likely that both versions will continue to operate simultaneously on networks well into the future. To date, most networks that use IPv6 support both IPv4 and IPv6 addresses.

The main characteristics of IPv4 and IPv6 can be summarized as follows. For more details, the reader is invited to the many widely available references on the subject.

IPv4

IPv4 uses 32-bit (4-byte) addresses in dotted decimal notation, for example 192.149.252.76, with each entry being a decimal digit – leading zeros can be omitted. An address is composed of a network and a host portion, which depend on the address class. Various address classes are defined: A, B, C, D, or E depending on initial few bits. The total number of IPv4 addresses is $2^{32} = 4,294,967,296$. NAT can be used to extend these address limitations.

The IPv4 header has a variable length of 20–60 bytes, depending on the selected IP options. The support of IPSec is optional and options are integrated in the header fields (before any transport header). The left-hand side of Figure 2.2 shows the structure of an IPv4 header.

IPv4 addresses are not associated with the concept of address lifetimes, unless the IP address has been assigned by a DHCP (for example, through a Wi-Fi access point).

IPv4 addresses are categorized into three basic types: unicast address, multicast address, and broadcast address. All IPv4 addresses are public, except for three address ranges that have been designated as private by IETF RFC 1918: 10.*.*.* (10/8), 172.16.0.0 through 172.31.255.255 (172.16/12) , and 192.168.*.* (192.168/16). Private address domains are commonly used within organizations. Private addresses cannot be routed across the Internet. IP addresses are assigned to hosts by DHCP or static configuration.

The typical minimum value of the maximum transmission unit (MTU) of a link – the maximum number of bytes that a particular link type supports – is 576 bytes.

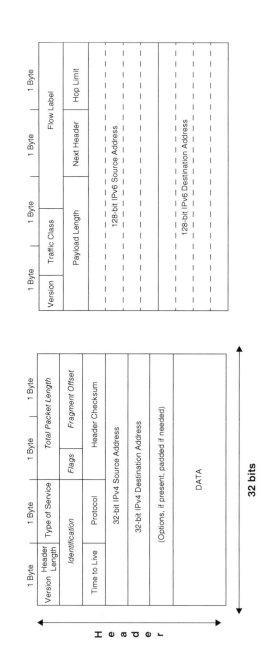

Figure 2.2 Header structure for IPv4 (left) and IPv6 (right) packets.

IPv6

IPv6 uses 128-bit (16-byte) addresses in hexadecimal notation (i.e., each entry corresponds to 4 bits), for example 3FFE:F200:0234: AB00. The basic architecture is 64 bits for the network number and 64 bits for the host number. Often, the host portion of an IPv6 address (or part of it) will be derived from a MAC address or other interface identifier. The total number of IPv6 addresses is $2^{128} = 3.4 \times 10^{38}$. There is no NAT support by design.

The IPv6 header has a fixed length of 40 bytes and there are no IP header options. IPSec needs to be supported and options are supported with extension headers (there is a simpler header format). The extension headers are AH and ESP (unchanged from IPv4), hop-by-hop, routing, fragment, and destination. The right-hand part of Figure 2.2 shows the structure of an IPv6 header.

IPv6 addresses have two lifetimes: preferred and valid. The preferred lifetime is always shorter than the valid one. After the preferred lifetime expires, the address must not be used as a source IP address to establish new connections if another preferred address exists. When the valid lifetime expires, the address is not used as a valid source or destination IP address. The rationale for a potentially finite lifetime of an IPv6 address is to match the disposable nature of things, such as a beacon with a two-year lifetime.

IPv6 addresses are categorized into three basic types: unicast, multicast, and anycast. IPv6 addresses can also be classified as either public or temporary (with limited lifetime). Temporary addresses can be routed globally, while IPv4 private addresses cannot. Temporary addresses are used for privacy purposes, concealing the actual identity of the initiator of a communication. Moreover, temporary addresses do not contain an interface identifier that is a MAC address. IPv6 addresses are self-assigned to hosts with stateless address auto-configuration or DHCPv6.

IPv6 has a lower boundary limit on MTU of 1280 bytes and therefore does not fragment packets below this threshold. Fragmentation and defragmentation at the link-layer must occur to transmit an IPv6 packet over a link of less than 1280 MTU, such as IEEE 802.15.4.

2.1.3   Transport Layer

The main goal of protocols at the transport layer is to provide host-to-host communication services for applications, hiding the underlying

TCP Segment Header Format

Bit #	0	7	8	15	16	23	24	31
0	Source Port				Destination Port			
32	Sequence Number							
64	Acknowledgement Number							
96	Data Offset	Res	Flags		Window Size			
128	Header and Data Checksum				Urgent Pointer			
160...	Options							

UDP Datagram Header Format

Bit #	0	7	8	15	16	23	24	31
0	Source Port				Destination Port			
32	Length				Header and Data Checksum			

Figure 2.3 Header structure for TCP (top) and UDP (bottom) datagrams.

networking strategy (dealt with at network layer) and specific link-by-link communication strategies (dealt with at link and physical layers).

2.1.3.1 TCP and UDP

The two widely used protocols at the transport layer are TCP and UDP: while TCP is used for connection-oriented transmissions, UDP is used for simpler messaging transmissions. In general terms, TCP creates virtual circuits between hosts and guarantees features such as reliability, flow control, congestion avoidance, and multiplexing, while UDP is a best-effort protocol that allows packet transmissions with minimum reliability (e.g., checking data integrity at the receiver. While TCP is the predominant transport layer protocol in the classical Internet, UDP, owing to its simplicity of implementation, is a very attractive option for IoT scenarios. TCP is suited for applications that require high reliability, and transmission time is relatively less critical. UDP is suitable for applications that need fast, yet "best-effort", transmission, such as games. UDP's stateless nature is also useful for servers that answer small queries from huge numbers of clients. This makes UDP a very attractive option for IoT scenarios.

TCP rearranges data packets in the order specified, whereas UDP has no inherent order: all packets are independent of each other. If ordering is required, it has to be managed by the application layer. TCP data is read as a byte stream, and no distinguishing indications are transmitted to signal message (segment) boundaries. In UDP, packets are sent individually and are checked for integrity only if they arrive.

TCP is "heavyweight", requiring three packets to set up a socket connection before any user data can be sent. It handles reliability and congestion control (in other words, TCP does flow control). UDP is "lightweight" – there is no ordering of messages, no tracking connections, and so on. In other words, UDP does not have an option for flow control.

In Figure 2.3, the header structure of TCP (top) and UDP (bottom) are shown: as expected, the TCP header is much larger than that for UDP. More precisely, each TCP header has ten required fields totaling 20 bytes (160 bits) and it can also optionally include an additional data section of up to 40 bytes in size; a UDP header contains 8 bytes.

The layout of TCP headers is as follows.

- Source TCP port (2 bytes) and destination TCP port (2 bytes) numbers are the communication endpoints for sending and receiving devices.

- Message senders use sequence numbers (4 bytes) to mark the ordering of a group of messages. Both senders and receivers use the acknowledgment number (4 bytes) field to communicate the sequence numbers of messages that are either recently received or expected to be sent.
- The data offset field (4 bits) stores the total size of a TCP header in multiples of 4 bytes. A header not using the optional TCP field has a data offset of five (representing 20 bytes), while a header using the maximum-sized optional field has a data offset of 15 (representing 60 bytes).
- Reserved data (3 bits) in TCP headers always has a value of zero. This field aligns the total header size as a multiple of four bytes (which is important for efficiency of data processing).
- TCP uses a set of six standard and three extended control flags (each an individual bit representing on or off, for a total of 9 bits) to manage data flow in specific situations. One bit flag, for example, initiates the TCP connection reset logic.
- TCP senders use a number called window size (2 bytes) to regulate how much data they send to a receiver before requiring an acknowledgment in return. If the window size becomes too small, network data transfer will be unnecessarily slow, while if the window size becomes too large, the network link can become saturated (unusable for any other applications) or the receiver may not be able to process incoming data quickly enough (also resulting in slow performance). Windowing algorithms built into the protocol dynamically calculate size values and use this field of the TCP header to coordinate changes between senders and receivers.
- The checksum value (2 bytes) inside a TCP header is generated by the protocol sender, and helps the receiver detect messages that are corrupted or tampered with.
- The urgent pointer field (2 bytes) is often set to zero and ignored, but in conjunction with one of the control flags, it can be used as a data offset to mark a subset of a message as requiring priority processing.
- Usage of optional TCP data (0–40 bytes) is beyond the scope of this chapter, but includes support for special acknowledgment and window-scaling algorithms.

The layout of UDP headers is as follows.

- Source UDP port (2 bytes) and destination UDP port (2 bytes) numbers are the communication endpoints for sending and receiving devices.

- The length field (2 bytes) in a UDP header represents the total size of each datagram, including both header and data. This field ranges in value from a minimum of 8 bytes (the required header size) to sizes above 65,000 bytes.
- Similar to TCP, a UDP checksum (2 bytes) allows receivers to cross-check incoming data for any corrupted bits.

2.1.4 Application Layer

Perhaps the most important reason for the widespread adoption of the Internet is the design and development of application layer protocols, which have been made available to developers as effective means to transmit information among endpoints of distributed applications without having to specify a custom communication protocol. This section will describe protocols that had an impact on the development of Internet-based applications and represent a valid introduction to IoT application scenarios.

2.1.4.1 HTTP

Hypertext Transfer Protocol (HTTP) [2, 3] is the most popular application layer protocol, and has been the true enabler of the Worldwide Web. Because of its extreme popularity, it is outside the scope of this book to present a detailed description of the protocol. We will outline the most prominent features in order to highlight why HTTP is perhaps the most important and widespread application layer on the Internet.

HTTP is a stateless text-based request/response protocol that defines the communication between a client and a server over a TCP connection. The HTTP protocol is characterized by the following features:

- URLs are used to give addresses to resources targeted by requests.
- Methods (or verbs) are defined in order to provide semantics at the application layer about the type of operation being executed by the server. The main methods used are typically GET, POST, PUT, and DELETE, although many other verbs have been defined, either in the HTTP specification itself or in subsequent updates to address specific issues.
- Header fields can be added to messages in order to convey further information, so that they are processed in the correct way and to provide additional semantics to communication.

- Status codes in responses provide a standard, uniform, and descriptive way to inform the client about the result of the request that has been served. Status codes are divided into classes:
  - **1xx** Informational
  - **2xx** Success
  - **3xx** Redirection
  - **4xx** Client error
  - **5xx** Server error

The solidity of the HTTP specification and its implementation in all programming languages and platforms make it common as an application-layer transport protocol for information transmitted between endpoints. Its use makes the design of distributed applications extremely efficient, because all efforts can be dedicated to defining the semantics of the data being exchanged.

In 2015, HTTP/2 [4] was released, with a particular focus on efficient communication and reduced consumption of network resources in terms of latency, header compression, and parallel flows over the same connection.

2.1.4.2 AMQP

The Advanced Message Queue Protocol (AMQP[1]) is a message-oriented protocol and a standard of the Organization for the Advancement of Structured Information Standards. The current version of AMQP is 1.0. AMQP can be seen as the asynchronous complement of HTTP.

AMQP is not an actual publish/subscribe (pubsub) protocol, but rather a specification for interoperable messaging for message-oriented middleware (MOM). AMQP defines a wire format; that is, the set of rules and principles that must be observed to create the stream of bytes to be transmitted. Therefore, any AMQP client can work with any MOM that implements AMQP.

AMQP is based on a queue server. Queues are message storage facilities. AMQP publishers can send messages directly to a queue or to an exchange, which is a topic-based router for message dispatching. Messages can be tagged with a routing key (a topic). Queues can be bound to an exchange for selective message dispatching to consumers. The routing key supports a dot-separated syntax that allows for different

---

1 `https://www.amqp.org`.

levels of expressiveness. Through the use of wildcards, a fine-grained routing of messages can be implemented.

Persistent queues store messages until a consumer pulls them off the queue; they are not tied to the existence of a consumer. Queues can, however, be dynamically created by a consumer and will then be destroyed when the consumer disconnects.

AMQP provides great flexibility in creating different messaging scenarios, such as message queue, fanout, routing, and even remote procedure calls (RPC). It supports reliable delivery of messages.

The most popular implementations of AMQP are RabbitMQ[2] and Apache ActiveMQ[3] .

2.1.4.3 SIP

The Session Initiation Protocol (SIP) [5] is an IETF standard application-layer control protocol that can be used to establish, modify, or terminate end-to-end sessions. SIP is a text-based client-server protocol, where the client sends SIP requests and the server responds to requests. The SIP architecture includes both end systems (terminals), also called SIP user agents, and intermediate systems, called SIP proxy, redirect or registrar servers, depending on their function. A "registrar server" is SIP server that receives registration requests issued by SIP user agents, and is used for maintaining the binding between the SIP user name (also called address-of-record; SIP AOR) and its current contact address, which can be used for reaching such user/resources. The mapping between SIP AORs and SIP contact URIs is called a location service and is an important component for resource discovery in SIP.

All SIP addresses are represented by URIs with the scheme sip:, and identify a name or contact address of a SIP user; a SIP user can be a real user, a system, an application, or some other kind of resource.

The proxy servers are intermediary entities that act as both server and client for making requests on behalf of other clients. A proxy server may act as an "outbound" proxy when used for routing SIP requests addressed to users that are not maintained in a local location service, or as a "far-end" (or "destination") proxy if the request is addressed to a user with an AOR maintained by the proxy and mapped to one or more SIP contact URIs.

---

2 https://www.rabbitmq.com/.
3 http://activemq.apache.org.

In contrast to proxy servers, redirect servers accept requests and replies to the client with a response message that typically provides a different contact address (URI) for the target of previous request.

SIP signaling between users consists of requests and responses. When a user agent (UA) wants to send a request to a remote user (identified by a SIP AOR), it may send the message directly to the IP address of the remote user's UA, or to the proxy server that is responsible for the target AOR (normally the fully qualified domain name of the proxy server is included in the AOR), or to a locally configured outbound proxy server. When the request reaches the target UA, the latter may optionally reply with some provisional 1xx responses and with one final response (codes 2xx for success, or 3xx, 4xx, 5xx and 6xx for failure).

SIP defines different request methods, such as INVITE, ACK, BYE, CANCEL, OPTIONS, REGISTER, SUBSCRIBE, and NOTIFY. When a UA wants to initiate a session it sends an INVITE message that may be responded to with provisional 1xx responses and a final response. The UA that issued the INVITE then has to confirm the final response with an ACK message. In contrast to all other SIP transactions, the INVITE transaction is a three-way handshake (INVITE/2xx/ACK).

Once the session is established, both endpoints (user agents) may modify the session with a new INVITE transaction, or tear-down the session with a BYE transaction (BYE/2xx). When the caller or the callee wish to terminate a call, they send a BYE request. SIP messages may contain a "body", which is treated as opaque payload by SIP.

Figure 2.4 shows an example of SIP message flow, including the registration of two UAs with their own registrar/proxy servers, and a session setup and tear-down from UA1 (identified by the SIP AOR sip:u1@P1) to UA2 (identified by the SIP AOR sip:u2@P2).

During an INVITE transaction, the SIP body is used to negotiate the session in terms of transport and application protocol, IP addresses and port number, payload formats, encryption algorithms and parameters, and so on. The negotiation follows an offer/answer paradigm: the offer is usually sent within the INVITE while the answer is in the 2xx final response. The most commonly used protocol for such negotiations is the Session Description Protocol (SDP), but other protocols may be used.

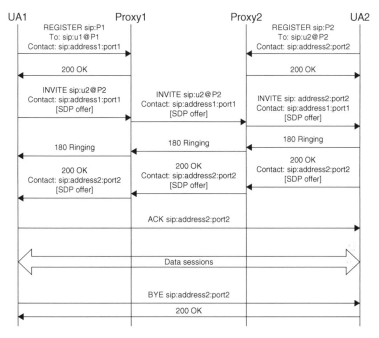

Figure 2.4 UA registration and session setup with two intermediate proxy servers.

2.2   The Internet of Things

The IoT and its applications are becoming a reality, with several commercial players developing innovative products in a variety of fields, such as home automation and smart cities. These products are starting to reach end users, who are now becoming aware of the integration of physical and cyber worlds. The forecast of billions (or trillions) of connected devices in future years is constantly being confirmed and can be objectively considered as a fact. The "gold rush" of the IoT era, driven on one hand by the will to demonstrate the feasibility of interconnecting everyday devices to people and, on the other hand, by the intention to make custom solutions into standards for public use, has created a plethora of closed vertical solutions. This is leading to a highly fragmented market, a babel of incompatible solutions, rather than a highly interoperable environment, which is what the Internet and the IoT should be like.

Current IoT applications consider personal devices to be gateways, which have to bridge the world of smart objects and the user. Typically, users must run custom software on their smartphones in order to interact with particular IoT objects. These apps have prior knowledge of the things they will work with: there are established communication protocols, data formats, and application-specific interaction patterns. In the long run, this approach represents a huge barrier to progress for a number of reasons.

- Smart objects manufactured by the same vendor typically need to be accessed through legacy software, resulting in a plethora of applications that end-users must install and use. This also has an impact on the creation of new applications for smart objects, whose interfaces (APIs) must be disclosed by the vendor to application developers.
- Mobility is a critical factor: people access services and use applications while on the go. This means they might enter environments that contain smart objects that they have not come across before (for example, because they enter a room in a hotel or museum they had never visited before, because some smart objects might have been deployed since their last visit, or because smart objects have changed their capabilities). While there are standard self-configuring mechanisms for discovering services and resources,(such as mDNS, DNS-SD, and web linking), these approaches are not adequate if users want to to fully and seamlessly interact with things.
- Smart objects should be able to adapt dynamically to particular conditions, such as a change in their battery level or hosted resources, or the presence of specific users. Smart objects should be able to drive different forms of interaction, which may not have been considered at the time of custom software implementation and would therefore require upgrades or new software to be installed.

2.2.1 Designing the Architecture of an IP-based Internet of Things

The IoT calls for adaptive interactions between humans and smart objects, with the goal of filling the gap between users, the physical world, and the cyber world. In order to avoid dependence on specific legacy software, this interaction should be driven by the smart objects that are pervasively deployed and accessed. All actors need to speak the same language: IP.

In order to prevent the fragmentation that results from vertical solutions, significant effort has been expended in private and public research groups and in standards organizations such as IEEE, IETF, and IPSO. This activity had two main goals:

- to define open standards for communication (e.g., 6LoWPAN/-CoAP);
- to map the traditional IP-based Internet stack to the IoT.

We can now assume that the IoT will be a network of heterogeneous interconnected devices. This will be the infrastructure for the so-called "Web of Things" (WoT). After being actively involved in this research and standardization phase, we feel that there is still a significant amount of work to be done to reach a state where the IoT can be accessed and exploited by end-users with the same simplicity that they experience on the web. At this point, it is time to really start interacting with smart objects, not just to communicate with them. The WoT is being designed around well-known concepts and practices derived from the Web, such as the REST paradigm. The REST paradigm was introduced to loosely couple client applications with the systems they interact with, to support long-term evolution of systems, and to provide robustness. While this proves to perfectly fit in machine-to-machine (M2M) scenarios, the loose coupling introduced at the application layer is not sufficient to enable the widespread adoption of applications that require humans to be in the loop, for example by providing input/output through some user interface. CoAP has been designated as the standard protocol for the WoT, similar to the position of HTTP for the web. In fact, CoAP has been designed to work with HTTP, to which it directly maps so as to give easy integration. While CoAP mainly targets M2M application scenarios, the widespread adoption of the WoT requires tools that allow humans to be in the loop. The Firefox web browser add-on Copper6 has been developed with the aim of narrowing the gap between the world of browsers and the world of things. Although this add-on provides a handy technical, debug console-like tool for working with CoAP-enabled smart objects, it does not really allow for easy interaction for any user. In our opinion, intuitive and easy-to-use interfaces for end-users are currently missing.

The evolution of mobile and wearable computing has changed the way people use online services; they are now always connected, whether at home or on the go. In this context, there is a concrete need to fill the gap between mobile devices and the IoT. A paradigm shift in areas such as human–computer interactions is needed to let people access and use the IoT with the same ease with which they can access the Internet and possibly also to enable new and more natural forms of interaction, which will widen the range of IoT users.

The evolution towards the IoT begins by rethinking and optimizing all of the relevant layers of the protocol stack. In the following, we outline the main characteristics of the communication protocols being considered for the IoT.

2.2.2 Physical/Link Layer

In this subsection, we focus on four communication protocol groups relevant for IoT scenarios, (mostly) at the physical/link layer:

- the IEEE 802.15.4 standard and Zigbee, which relies on IEEE 802.15.4 and introduces application management on top of this (i.e., Zigbee is not strictly at the physical/link layer)
- low-power Wi-Fi, an amendment of the Wi-Fi protocol family that is attractive for IoT applications
- Bluetooth and its newest energy-efficient version Bluetooth Low Energy (BLE), which are extremely attractive for IoT because of their widespread availability, notably in almost every smartphone
- protocols in the area of power line communications (PLC).

While the first three groups are relevant for wireless communications, the last one is used in wired communications (in electrical cables).

2.2.2.1   IEEE 802.15.4 and ZigBee

The IEEE 802.15 working group defined the physical layer (PHY) and the medium access sub-layer (MAC) for low-complexity, low-power-consumption, low-bit-rate WPAN connectivity. The IEEE 802.15.4 standard, approved in 2003 and amended several times in the following years, contributes to all of these aims, and several compliant products are already available on the market, even if more as development kits than as real end-products.

The physical layer of IEEE 802.15.4 foresees the use of one of three possible unlicensed frequency bands:

- 868.0–868.6 MHz: used in Europe, and allows one communication channel (versions published in 2003, 2006, 2011);

- 902–928 MHz: used in North America, up to ten channels (2003), extended to thirty (2006);
- 2400–2483.5 MHz: used worldwide, with up to sixteen channels (2003, 2006).

The original 2003 version of the standard specifies two physical layers based on direct sequence spread spectrum (DSSS) techniques: one working in the 868/915 MHz bands with transfer rates of 20 and 40 kbit/s, and one in the 2450 MHz band with a rate of 250 kbit/s.

The 2006 revision (IEEE 802.15.4b) improves the maximum data rates of the 868/915 MHz bands, so they support 100 and 250 kbit/s. It goes on to define four physical layers depending on the modulation method used. Three of these preserve the DSSS approach:

- In the 868/915 MHz bands, binary or offset quadrature phase shift keying can be used (the latter being optional).
- In the 2450 MHz band, using the latter.
- An alternative, optional 868/915 MHz layer is defined using a combination of binary keying and amplitude shift keying; it is thus based on parallel sequence spread spectrum (PSSS).

Dynamic switching between supported 868/915 MHz PHYs is possible.

Beyond these three bands, the IEEE 802.15.4c study group considered the newly opened 314–316 MHz, 430–434 MHz, and 779–787 MHz bands in China, while the IEEE 802.15 Task Group 4d defined an amendment to 802.15.4-2006 to support the new 950–956 MHz band in Japan. The first standard amendments by these groups were released in April 2009.

In August 2007, IEEE 802.15.4a was released. This expanded the four PHYs available in the earlier 2006 version to six, including one PHY using direct sequence ultra-wideband and another using chirp spread spectrum. The ultra-wideband PHY is allocated frequencies in three ranges: below 1 GHz, between 3 and 5 GHz, and between 6 and 10 GHz. The chirp spread spectrum PHY is allocated spectrum in the 2450 MHz ISM band.

In April 2009, IEEE 802.15.4c and IEEE 802.15.4d were released. These expanded the available PHYs through addition of several new PHYs: one for the 780 MHz band using O-QPSK or MPSK and another for the 950 MHz band using GFSK or BPSK.

IEEE 802.15.4e, finally approved in 2012, was defines a MAC amendment to the existing 802.15.4-2006 standard. It adopts a channel-hopping strategy to improve support for the industrial markets, and increases robustness against external interference and persistent multi-path fading.

The IEEE 802.15.4 family provides for low bit-rate connectivity in the personal operating space; typically between 10 and 100 m. Full support of mesh networks for battery-powered nodes is provided, through the classification of devices into two different types: full function devices (FFD) and reduced function devices (RFD). An IEEE 802.15.4 network should include at least one FFD operating as the PAN coordinator for special (but not centralized) functions, whereas all the other FFDs should make up the wireless sensor network (WSN) backbone; RFDs, which are usually intended as the "leaf" nodes of the WSN spanning tree, perform simple tasks more related to sensing than networking. An RFD can communicate only with one FFD, while an FFD can communicate with both RFDs and FFDs.

In the late 2000s, on the basis of the current IEEE 802.15.4 specifications, a consortium of hundreds of companies agreed the adoption of an industrial standard called ZigBee; the name was inspired by the social behavior of bees, which work together to tackle complex tasks. ZigBee exploits cooperation to allow for the multi-hop exchange of messages and it adds logical network, security, and application management functions on top of the referenced IEEE 802.15.4 standard by defining the upper layers of the protocol stack, from network to application. In addition, ZigBee defines application profiles, a set of template-based descriptions of device configurations, each one specialized for work in a common cooperative and distributed application. Aside from its technical aspects, one of the main tasks of the ZigBee Alliance is to ensure interoperability among devices made by different manufacturers, thus expanding their potential adoption.

2.2.2.2 Low-power Wi-Fi

Low-power Wi-Fi refers to IEEE 802.11ah, an amendment of the IEEE 802.11-2007 wireless networking standard, called Wi-Fi HaLow by the Wi-Fi Alliance. Wi-Fi HaLow is expected to enable a variety of new power-efficient use cases in smart homes, connected cars, and digital healthcare, as well as industrial, retail, agriculture, and smart city environments. Wi-Fi HaLow extends Wi-Fi into the 900 MHz band, enabling the low-power connectivity necessary for the creation of large groups of stations or sensors that cooperate to share signals, supporting the concept of the IoT. Wi-Fi HaLow's range is nearly twice that of today's Wi-Fi, and will not only be capable of transmitting signals further, but will also provide more robust connections in challenging environments and where the ability to penetrate walls or other barriers is an important consideration.

802.11ah could solve many of the problems with deploying large-footprint Wi-Fi networks by allowing for a significant number of devices, providing power-saving services and allowing for long distances to the AP. A typical 802.11ah AP could associate more than 8,000 devices in a hierarchical ID structure within a range of 1 km, making it ideal for areas with a high concentrations of sensors and other small devices, such as street lamp controllers and smart parking meters. The 802.11ah standard also includes new PHY and MAC layers, grouping devices into traffic indication maps to accommodate small units (such as sensors) and M2M communications.

2.2.2.3 Bluetooth and BLE

Bluetooth is a standard wire-replacement communications protocol primarily designed for low-power consumption and short communication ranges. The transmission range is power dependent. The specifications were formalized by the Bluetooth Special Interest Group (SIG). The SIG was formally established by Ericsson, IBM, Intel, Toshiba and Nokia in 1998: today it has a membership of over 30,000 companies worldwide. While Bluetooth 3.0, introduced in 2009, supported a data rate of 25 Mbit/s with a transmission range of 10 m, with the latest Bluetooth 5.0, introduced in 2016, the data rate and transmission range have increased to 50 Mbit/s and 240 m. On top of the physical layer, link-layer services including medium access, connection establishment, error control, and flow control are provided. The upper logical link control and adaptation protocol provides multiplexing for data channels, fragmentation and reassembly of larger packets. The other upper layers are the Generic Attribute Protocol, which provides for efficient data collection from sensors, and the generic access profile, which allows for configuration and operation in different modes, such as advertising or scanning, and connection initiation and management.

The Bluetooth Core Specification version 4.0 (known also as "Bluetooth Smart") was adopted in 2010. Bluetooth 4.0 includes classic Bluetooth, Bluetooth High Speed and Bluetooth Low Energy (BLE) protocols. Bluetooth High Speed is based on Wi-Fi, while classic Bluetooth consists of legacy Bluetooth protocols. BLE, previously known as Wibree, is a subset of Bluetooth 4.0 with an entirely new protocol stack for rapid build-up of simple links. It is aimed at very low power applications running off a coin cell battery. Chip designs allow for two types of implementation: dual-mode and single-mode.

Starting from version 4.2, IoT-oriented features have been introduced into Bluetooth:

- low energy secure connection with data packet length extension (v4.2);
- link layer privacy (v4.2);
- IP support profile (v6.0)
- readiness for Bluetooth Smart Things to support connected homes (v4.2);
- connectionless services, such as location-relevant navigation of low-energy Bluetooth connections (v5.0).

BLE uses a short-range radio with minimal power use, which can operate for a much longer time (even for years) compared to previous versions. Its range coverage (about 100 m) is ten times that of the classic Bluetooth while its latency is 15 times shorter. BLE can be operated using a transmission power of between 0.01 and 10 mW. With these characteristics, BLE is a good candidate for IoT applications. The BLE standard has been developed rapidly by smartphone makers and is now available in most smartphone models. The feasibility of using this standard has been demonstrated in vehicle-to-vehicle communications as well as in WSNs.

Compared to ZigBee, BLE is more efficient in terms of energy consumption and the ratio of transmission energy per transmitted bit. BLE allows devices to operate as masters or slaves in a star topology. For the discovery mechanism, slaves send advertisements over one or more dedicated advertisement channels. To be discovered as a slave, these channels are scanned by the master. When they are not exchanging data, the devices are in sleep mode.

2.2.2.4 Powerline Communications

Power line communications (PLC) involve the use of existing electrical cables to transport data and have been investigated for a long time. Power utilities have been using this technology for many years to send or receive (limited amounts of) data on the existing power grid. Today, PLCs represent a very appealing area of application for IoT technologies.

Relevant IoT-oriented smart grid communication protocols can be summarized as follows.

- *PRIME*: intended for PLC-based modems operating in the frequency range between 42 kHz and 88 kHz using orthogonal frequency-division multiplexing;
- *HomePlug*: operating at frequencies of up to 400 kHz;
- *G3-PLC*: intended for PLC-based modems operating in a sub-frequency range of the CENELEC A band, from 35 to 91 kHz;
- *G.hnem*: the specification was drafted by ITU and selected G3-PLC and PRIME as annexes to its main body.
- *IEEE P1901.2*: defined by IEEE and adopting G3- PLC and PRIME.

2.2.3   Network Layer

The envisioned IP-based IoT will consist of trillions of connected devices. This unprecedented magnitude demands strategic choices to be taken in the design of a global network of smart objects. On the one hand, there is the need to address each and every smart object individually (and globally). On the other hand, the use of IPv4 cannot be a long-term approach. The depletion of IPv4 addresses makes it impossible to assign public IPv4 addresses to objects. The use of IPv4 would require the introduction of NAT techniques in order to provide an extended address space to smart objects. This would involve complex configuration management to ensure reachability of smart objects. As a result, the scalability, manageability, and ease of deployment of smart objects would be in jeopardy.

Stemming from these considerations, the only feasible solution to create a global, sustainable, and scalable IoT is to adopt IPv6 at the network layer. In particular, IPv6 provides some beneficial features that make its adoption convenient:

- its addresses are 128 bits long, thus making it possible to assign about 3.4×10^{38} unique IP addresses;
- it integrates IPSec for security;
- it provides link-local and global addresses, derived from the device's MAC address, with a prefix of fe80::/10, or provided by the network's router.

2.2.3.1 The 6LoWPAN Adaptation Layer

Low-power wireless personal area networks (WPANs) have special characteristics, that set them apart from earlier link-layer technologies. These include limited packet size (a maximum 127 bytes for IEEE 802.15.4), various address lengths, and low bandwidth. These characteristics necessitate an adaptation layer that fits IPv6 packets to the IEEE 802.15.4 specifications. The IETF 6LoWPAN working group developed such a standard in 2007. 6LoWPAN is the specification of mapping services required to maintain an IPv6 network over low-power WPANs. The standard provides header compression to reduce the transmission overhead, fragmentation to meet the IPv6 MTU requirement, and forwarding to link-layer to support multi-hop delivery. In general, the goal of 6LoWPAN is to transmit a small IPv6 datagram over a single IEEE 802.15.4 hop.

2.2.4 Transport Layer

IoT scenarios typically call for energy-efficient, lightweight, and non CPU-intensive approaches to communication. This is a result of the limited capabilities of smart objects. For these reasons, UDP is the typical choice for transport-layer communication in the IoT. Of course, this choice results in the impossibility of enjoying the nice features that TCP provides, such as retransmission, ordering, and congestion control. These must be implemented in a higher layer if needed by an application.

There are other transport-layer protocols that may be considered, such as SCTP [6], which focuses on stream control. However, there has not been a concrete effort by the research community, academia, or industry to define an IoT-oriented transport-layer protocol as has been done for all other layers.

2.2.5 Application Layer

The experience gained with the Internet and the importance attributed to application-layer protocols have been critical to the awareness that the IoT needs a dedicated application-layer, which must take into account all the requirements of, and the conditions deriving from, lower layers. A dedicated web transfer protocol for low-power and lossy networks (LLNs), called the Constrained Application Protocol (CoAP), has been defined, with the aim of replicating the outstanding experience of HTTP and the resulting widespread adoption of the web. Other application-layer protocols have also been proposed in order to provide alternatives to the stateless request/response communication paradigm offered by CoAP. Examples include pub/sub communications (an example of which is the MQTT protocol), and introducing sessions such as CoSIP.

In this section, we will provide a detailed presentation of CoAP and CoSIP in order to highlight their features, the design choices behind them, and how they can represent the enablers for the development and deployment of efficient, complex, large-scale IoT applications.

2.2.5.1 CoAP

LoWPANs are typically characterized by:

- *Small packet size*: Since the maximum physical layer packet for IEEE 802.15.4 is 127 bytes, the resulting maximum frame size at the MAC layer is 102 octets. Link-layer security imposes a further overhead, which, in the maximum case (21 octets of overhead in the AES-CCM-128 case, versus 9 and 13 for AES-CCM-32 and AES-CCM-64, respectively), leaves only 81 octets for data packets.
- *Low bandwidth*: The limited bandwidth does not allow for data to be transferred at high rates (data rates are of the order of tens or a few hundreds of kilobits per second). It is important to exchange as little data as possible to minimize latency of transmission.
- *Low power*: Some devices are battery operated, so energy consumption is a critical issue. Since radio-related operations are the most energy-consuming, it is desirable to minimize the amount of data to be transferred to keep radio utilization as low as possible.
- *Low cost*: Devices are typically associated with sensors, switches, and so on. This drives some of their other characteristics such as low processing power, low memory, and so on. Numerical values for "low" are not given on purpose: costs tend to change over time
- *Unreliability*: Devices are subject to uncertain radio connectivity, battery drain, device lockups, and physical tampering.
- *Duty cycling*. Devices connected to a LoWPAN may sleep for long periods of time in order to save energy, and are thus unable to communicate during these sleep periods.

Adopting IPv6 in constrained environments through 6LoWPAN introduces:

- packet fragmentation (into small link-layer frames) and compression at the transmitter;
- fragment reassembly (from small link-layer frames) and decompression at the receiver.

In 6LoWPANs, fragmentation can increase the probability of packet delivery failure. Given the small packet size of LoWPANs, applications must send small amounts of data:

- less data $\Rightarrow$ fewer fragments to be sent $\Rightarrow$ lower energy consumption (CPU/TX/RX);
- less data $\Rightarrow$ fewer fragments to be sent $\Rightarrow$ lower packet loss probability.

Due to the limited capabilities of objects in the IoT, the distinction among the different layers of the protocol stack is not as strong as in the traditional Internet. This means that design choices taken at a particular layer may (and typically do) have an impact on all the other layers. A *cross-layer design* is therefore required to ensure that things work as expected and to prevent a bad design from affecting the functioning of the object. For instance, HTTP and TCP are not well suited in this area due to overhead they introduce. On the one hand, HTTP is a text-based and very verbose protocol. On the other hand, TCP is a connection-oriented transport and requires setting up and maintaining of connections, which is burdensome both from a processing and a communication perspective. Moreover, duty-cycling is not compatible with keeping connections alive.

For all the above reasons, application-layer protocols for the IoT must be designed carefully in order to take into account the constraints deriving from lower layers and the very nature of smart objects.

The IETF candidate as application-layer protocol for the IoT is the Constrained Application Protocol (CoAP), defined in RFC 7252 [7]. CoAP was designed by the IETF Constrained RESTful Environments (CoRE) working group and became an Internet standard in mid-2014. CoAP is a lightweight application-layer protocol purposedly designed to bring web functionalities to constrained devices that operate in LLNs.

In order to meet the requirements of constrained environments, CoAP differs from HTTP in a number of ways. First, it uses UDP as the underlying transport protocol instead of TCP, thus removing the burden of establishing and maintaining connections, which may be infeasible for smart objects that have limited capabilities and may be duty-cycled. As a consequence, CoAP implements its own reliability mechanisms for message retransmission, which cannot be guaranteed by the transport-layer protocol. In addition, in order to minimize overhead, and unlike HTTP, CoAP uses a binary format instead of a text-based format.

CoAP has been designed to bring the REST paradigm (see Section 3.3) to the IoT. CoAP implements a request/response (client/server) communication model on top of UDP – or its secure version, Datagram Transport Layer Security (DTLS) — supporting four basic methods: POST, GET, PUT, and DELETE.

Protocol overview

CoAP is an application-layer protocol designed to be used by constrained devices in terms of computational capabilities, which may feature limited battery and operate in constrained (low-power and lossy) networks.

CoAP is a RESTful protocol According to the REST paradigm, CoAP URIs identify the resources of the application. A resource representation is the current or intended state of a resource referred to the server through a proper namespace. Representations of resources are exchanged between a client and a server. A client that is interested in the state of a resource sends a request to the server; the server then responds with the current representation of the resource.

CoAP maps to HTTP CoAP maps to HTTP easily in order to guarantee full integration with the web. The CoAP protocol stack mirrors HTTP's, as shown in Figure 2.5. The mapping allows protocol translation to be performed easily on dedicated proxies in order to guarantee full interoperability between HTTP clients and CoAP servers and vice versa. This mechanism enables CoAP-unaware clients, such as legacy

Figure 2.5 CoAP protocol stack vs. HTTP protocol stack.

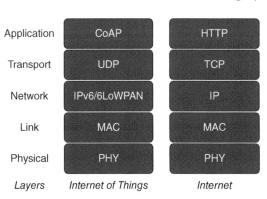

Layers	Internet of Things	Internet
Application	CoAP	HTTP
Transport	UDP	TCP
Network	IPv6/6LoWPAN	IP
Link	MAC	MAC
Physical	PHY	PHY

HTTP-based applications, to work with smart objects without requiring any changes.

CoAP is a binary protocol CoAP minimizes overhead by adopting a binary message format. CoAP messages are formed from some mandatory fields (Version, Type, Token, Code, Message-ID) and other optional fields (CoAP Options and Payload). The syntax of CoAP messages has been designed not only to keep messages small, but also to make them extremely easy and light to parse, so that smart objects can receive a benefit in terms of energy consumption.

CoAP runs on top of UDP CoAP runs on top of UDP, which is the most lightweight transport in terms of overhead (only 8 bytes are added by the header). Moreover, the connectionless nature of UDP does not introduce communication, processing, or memory overheads due to the establishment and maintenance of TCP connections.

CoAP embeds IoT-oriented features Due to the nature of the operational scenarios in which it is going to be used, CoAP introduces some IoT-oriented features, such as:

- resource observation
- asynchronous message exchange
- multicast communication.

CoAP URI

CoAP URIs use either the *coap*: or *coaps*: scheme (Figure 2.6), the latter referring to secure CoAP (CoAPs). CoAPs uses DTLS as a secure transport, similar to HTTP's use of TLS. The default UDP ports are 5683 for *coap*: and 5684 for *coaps*:.

CoAP URI

coap://	example.com	:5683	/sensor	?id=1
scheme	host	port	path	query
coaps://	example.com	:5684	/sensor	?id=1

Figure 2.6 CoAP URI.

CoAP messaging

CoAP is based on a request/response communication model. Since CoAP is based on UDP, reliability is not provided by the transport layer. Retransmission may be therefore needed, especially in LoW-PAN, where the physical conditions and radio protocols used may have detrimental effects on communications. CoAP implements reliability internally, and handles these situations by implementing a retransmission and deduplication mechanism.

The connectionless nature of UDP results in an asynchronous communications model. This means that, unlike connection-based communications, request and responses must be matched. CoAP includes fields used for request/response matching.

CoAP is characterized by two layers, as shown in Figure 2.7:

- The *messaging layer* provides support for duplicate detection and (optionally) reliability.
- The *requests/responses* layer provides support for RESTful interactions, through GET, POST, PUT, DELETE methods.

CoAP messages are exchanged between UDP endpoints (the default port is 5683). CoAP messages, as in HTTP, are either requests or responses. All messages include a 16-bit message ID used to detect duplicates and for optional reliability (acknowledgements of received

Figure 2.7 CoAP messaging model.

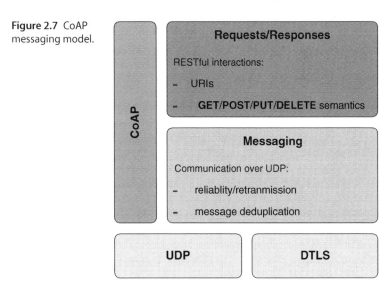

messages). In order to match requests and responses, an 8-bit token field is used. Message IDs and tokens are separate concepts and they actually work "orthogonally". The former is used to detect duplicates of a message due to retransmission that are received by an endpoint; in other words, no two messages can have the same message ID. The latter is used to link and match requests and responses; a response must report the same token as its corresponding request.

Reliability The use of UDP, especially in LLNs, might lead to lost messages. Reliability cannot be provided by transport layer (as it would be if TCP were used) and must be ensured at application layer. The definition of CoAP over other reliable transports, such as TCP, TLS, and WebSockets, is a work in progress [8]. In CoAP, reliability is implemented by performing retransmission with exponential back-off (timeout is doubled at each retransmission). If the sender of a message does not receive an expected acknowledgement, it resends the message (using the same message ID). A maximum of four retransmissions can be performed.

CoAP defines four types of message:

- CON (confirmable): used for messages that must be transmitted reliably;
- NON (non-confirmable): used for messages for which reliability is not needed;
- ACK (acknowledgment): used to acknowledge the reception of a CON message;
- RST (reset): used to cancel a reliable transmission.

Reliability is achieved thanks to the following rules (also shown in Figure 2.8):

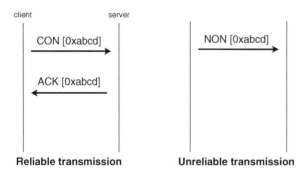

Figure 2.8 CoAP reliability.

- When a CON request is sent, an ACK response is required to be sent back to acknowledge the correct reception of the request.
- When a NON request is sent, the response should be returned in a NON message.
- CON messages provide reliability: if no ACK response is returned prior to a retransmission timeout (with exponential back-off), the message is retransmitted.

Piggy-backed and separate responses If a CON CoAP request is sent, CoAP responses can be either: *piggy-backed*, if the server can respond immediately, or *separate*, if the server cannot respond immediately.

With piggy-backed responses, the responses are included in the ACK message that acknowledges the request (implicit acknowledgement). With separate responses, the server first sends an ACK message to acknowledge the request; when the data is available, the server sends a CON message containing the response. This is acknowledged by the client with a new ACK message. The two behaviors are shown in Figure 2.9.

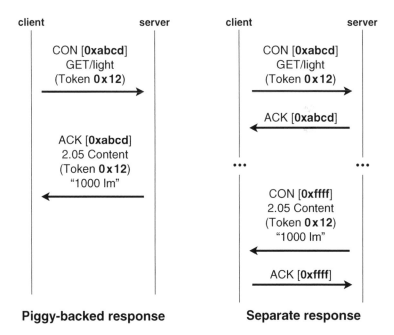

Piggy-backed response   **Separate response**

Figure 2.9 Piggyback response and separate responses.

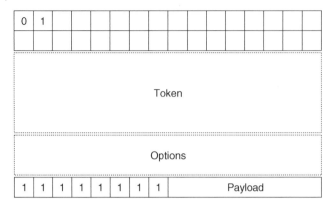

Figure 2.10 CoAP general message format.

CoAP message format

The general structure of a CoAP message is shown in Figure 2.10. A CoAP message is formed by a header and, optionally, a payload. The header's structure is detailed in Figure 2.11.

Version Ver (2 bits) specifies the protocol version number. The current version of CoAP sets these bits to 01 mandatorily.

Type T (2 bits) specifies the type of message. The following values are defined:

0 = CON is used for confirmable messages
1 = NON is used for non-confirmable messages
2 = ACK is used for acknowledgment messages
3 = RST is used for reset messages.

Token length TKL (4 bits) is the length in bytes of the Token field. Valid values are 0 to 8, while values 9–15 are reserved. A value of 0 indicates that the message will not include a token.

Code Code (8 bits) describes the message. The code is divided into a 3-bit *class* and a 5-bit *meaning*. The following values are defined for the code:

- Class 0 (000) identifies a request message.
- Class 2 (010) identifies a "Success" response.

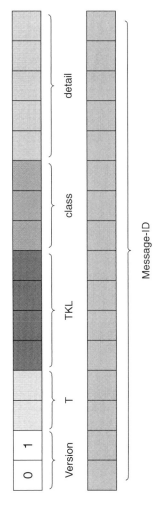

Figure 2.11 CoAP message header detail.

- Class 4 (100) identifies a "Client error" response (the client should not repeat the request as it will fail again).
- Class 5 (101) identifies a "Server error" response (the request failed due to a server error; the client might repeat the request, depending on the failure reason reported by the server).

A simple check on the presence of any bit set to 1 can be performed to determine immediately whether the message is a request or a response. Similarly, a check on the first bit can be performed to determine immediately whether the response was successful or not. Finally, a check on last bit can be performed to determine whether a response refers to a client or a server error. The structure of the code is an example of how CoAP has been designed in order to simplify parsing operations and to take processing load away from smart objects.

For requests, the detail is used to indicate the method of the request:

1 = GET
2 = POST
3 = PUT
4 = DELETE.

For responses, the detail is used to give additional information related to the response. Some examples of response codes are:

- Success responses: 2.01 Created, 2.02 Deleted, 2.04 Changed, 2.05 Content
- Client error responses: 4.00 Bad Request, 4.02 Bad Option, 4.04 Not Found, 4.05 Method Not Allowed, …
- Server error responses: 5.00 Internal Server Error, 5.01 Not Implemented, …

Message-ID Message-ID (16 bits) is the message identifier used to detect duplicates and for optional reliability.

Token (optional) Token (TKL bits) is the message token. It is used to match responses to requests independently from the underlying messages. The length in bytes is specified in the TKL field. This field is optional. If not used, TKL is 0.

CoAP options CoAP request and response semantics are carried in CoAP messages. Required and optional information for all messages, such as the URI and payload media type, are carried as *CoAP options.* Options have the same role as in HTTP headers. Table 2.1 shows all the options defined in the CoAP specification.

CoAP defines a number of options that can be included in a message. These options are encoded as TLV (type-length-value). Each option specifies the *option number* of the CoAP option (registered on the IANA registry), the *option length*, and the *option value.*

In order to maximize compactness, options are encoded in a very size-efficient way, called *delta encoding.* According to delta encoding, options are sorted in ascending order of their number. Instead of the actual option number, an *option delta* (4 bits) is reported; that is, the option number is calculated as a delta from the preceding option. The option number can be calculated by simply summing all the option delta values of the current option and all previous ones. Since the option delta is 4 bits long, the maximum delta between two consecutive options is 15 at most. In order to support bigger deltas

Table 2.1 CoAP options.

Option number	Option name	Format	Length (bits)
1	If-Match	opaque	0–8
3	Uri-Host	string	1–255
4	ETag	opaque	1–8
5	If-None-Match	empty	0
7	Uri-Port	uint	0–2
8	Location-Path	string	0–255
11	Uri-Path	string	0–255
12	Content-Format	uint	0–2
14	Max-Age	uint	0–4
15	Uri-Query	string	0–255
17	Accept	uint	0–2
20	Location-Query	string	0–255
35	Proxy-Uri	string	1–1034
39	Proxy-Scheme	string	1–255
60	Size1	uint	0–4

between options, an *extended option delta* is used. An extended delta adds 1 or 2 bytes after the option length, depending on the value of the option delta. The option delta can have the following values:

- 0 to 12 (no extended option);
- 13 (extended option): the value of the option ranges from 13 to 268 – 1 byte for extended delta is added after option length;
- 14 (extended option): the value of the option ranges from 269 to 65804 – 2 bytes for extended are added delta after option length.

If option delta is 13, 1 byte is used for the extended delta. The actual option number is calculated as (13 + value of extended delta). Since there is 1 byte added for the delta, the maximum option number is 13 + 255 = 268. If the option delta is 14, 2 bytes are used for the extended delta. The actual option number is calculated as (269 + value of extended delta). Since there are 2 bytes added for the delta, the maximum option number is 268 + 65,536 = 65,804. The use of the extended delta is shown in Figure 2.12.

After the option number, the option length (4 bits) is reported. Similar to the option number, the option length can be 15 at maximum. In order to allow for larger values, an *extended option length* is used. An extended length adds 1 or 2 bytes after the option length or after the extended delta (if present), depending on the value of the option length. The option length can have the following values:

- 0 to 12 (no extended option);
- 13 (extended option): 1 byte for extended length after option length or after extended delta;
- 14 (extended option): 2 bytes for extended length after option length or after extended delta.

If the option length is 13, with 1 byte for extended length, the option length is calculated as (13 + value of extended delta). If option length

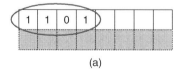

(a)

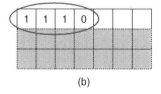

(b)

Figure 2.12 CoAP extended option delta with (a) 1 byte and (b) 2 bytes added after the option length.

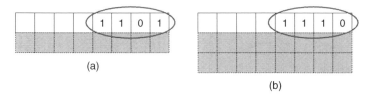

Figure 2.13 CoAP extended option length with (a) 1 byte and (b) 2 bytes added after the option length or extended delta.

is 14, with 2 bytes for extended length, the option length is calculated as (269 + value of extended delta). The use of extended length is shown in Figure 2.13.

Options can be either repeatable or not. Repeatable options report a delta of 0.

The Uri-Host, Uri-Port, Uri-Path, and Uri-Query options are used to specify the target resource of a request to a CoAP origin server. Uri-Path and Uri-Query options are repeatable. The Uri-Host option specifies the Internet host of the resource being requested. The Uri-Port option specifies the transport layer port number of the resource. Each Uri-Path option specifies one segment (segments are separated by a slash) of the absolute path to the resource. Each Uri-Query option specifies one argument parameterizing the resource. Figure 2.14 shows how to proceed to create the CoAP options for the CoAP URI `coap://example.com/people/123`.

The Location-Path and Location-Query options together indicate a relative URI that consists either of an absolute path, a query string or both. A combination of these options is included in a 2.01 (Created) response to indicate the location of the resource created as the result of a POST request.

The Content-Format option indicates the format of the representation included in the payload. The Accept option can be used to indicate which Content-Format is acceptable to the client. The representation format is a content format identifier; that is, a number defined in the CoAP Content Format Registry. Table 2.2 reports the defined content format identifiers.

Resource observation

The state of a resource on a CoAP server can change over time. Clients can get the most recent state in one of two ways: polling or observing. Polling means performing GET requests periodically. This

Option Name	Option Number	Option Delta	Option Value	Option Length (bytes)
Uri-Host	3	3	example.com	11
Uri-Path	11	8	people	6
Uri-Path	11	0	123	3

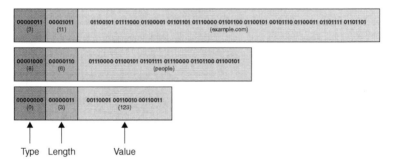

Type   Length   Value

Figure 2.14 Procedure to construct CoAP message options to target URI `coap://example.com/people/123`.

Table 2.2 CoAP content formats.

Media type	Encoding	Content format identifier
text/plain	charset=utf-8	0
application/link-format	-	40
application/xml	-	41
application/octet-stream	-	42
application/exi	-	47
application/json	-	50

Observer   Subject

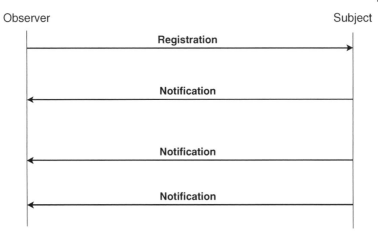

Figure 2.15 Observer pattern.

is a typical HTTP pattern. However, polling is inefficient for a number of reasons. Resources may not change state between two successive GET requests, thus making some of these requests unnecessary. Since all requests consume energy in the constrained CoAP server, this approach can have negative effects on battery-operated objects. Moreover, if the resource changes its state multiple times between successive GET requests, for some time the client was not aware of the most up-to-date state.

The Observe option [9] was introduced to avoid polling (RFC 7641) and to create a more suitable pattern for constrained environments. The Observe option implements an observer design pattern (Figure 2.15): when included in a GET request, the server is instructed to send notifications to the client whenever the state of the resource changes. This is similar to a pub/sub model, but is intrinsically very different. The communication is still based on requests and responses (rather than subscriptions and publishers), but multiple responses are sent after a single request. Responses are linked to the request because they report the same token.

The Observe option has been assigned the option number 6. Its length can be between 0 and 3 bytes. The option value is an unsigned integer:

- 0 is used for registration (start observing);
- 1 is used for deregistration (stop observing);

- other values are used to sort notifications (since UDP is used, notifications may arrive in the wrong order) or to ignore older notifications if a more up-to-date state has been notified.

Figure 2.16 shows a typical resource observation. The client performs a GET a request for a specific resource and contextually registers itself to the server to receive updates related to resource changes by adding an Observe option with value 0 and a specific token. The server replies with a 2.05 response with the same token and an Observe option value including the representation of the resource in the payload. Whenever the state of the resource changes, the server sends another response to the client with the same token and a different (subsequent) Observe option value. When the client stops being interested in updates on the resource, it sends another GET request to the server with the same token but with an Observe option value of 1. The server will reply with a final response and will no longer send updates.

The Observe option requires the server to maintain a list of registered clients (endpoint and token) for each resource. This can introduce an overhead on the smart object. Resources are therefore not necessarily observable. The server can mark a resource as observable in the CoRE Link Format representation of the link, using an obs attribute.

Blockwise transfers

CoAP was designed to achieve minimal overhead. In fact, CoAP messages work well for small payloads. CoAP messages are encapsulated in UDP packets, which have a maximum size of 65536 bytes. However, it might happen that applications need to transfer larger payloads. In other cases, constrained servers or clients may not be able to process payloads of arbitrary sizes. Let us think about a smart object with a buffer that limits it to accept payloads of 128 bytes, but where 1 kilobyte of data needs to be transferred. Blockwise transfers have been purposely defined to divide "large" amounts of data into several chunks of a given size [10]. The meaning of the term "large" depends on the specific scenario: it may mean larger than a UDP packet or larger than the maximum size of an object's buffer.

RFC 7959 defines two new options called Block1 and Block2 for this goal. Block1 and Block2 options provide a minimal way to transfer larger representations in a blockwise fashion. The two options are used to convey the necessary information to split a large payload into smaller chunks and to control the transfer of each chunk in such a way that it can be successfully reconstructed at the other endpoint.

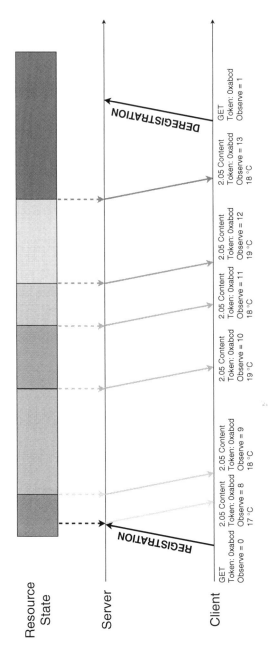

Figure 2.16 CoAP resource observation.

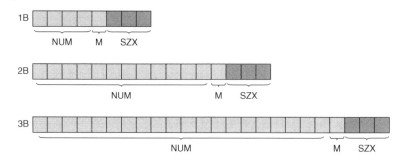

Figure 2.17 Anatomy of a Block option value.

Block1 (option number 27) is related to the request payload and is typically used in conjunction with POST/PUT requests. Block2 (option number 23) is related to the response payload and is typically used in conjunction with GET requests. Clients and servers can also negotiate a block size that can be used for all transfers, using the Size1 and Size2 options.

The value of the Block option is variable in size (0 to 3 bytes) and is formed by the following three fields, as shown in Figure 2.17:

- NUM: the relative number of the block within a sequence of blocks with a given size.
- M (1 bit): a "more" bit, indicating whether more blocks will follow or this is the last block.
- SZX (3 bits): the size exponent of the block; SZX can be a number from 0 to 6, while the value is reserved and cannot be used. The actual size of the block is computed as 2^{4+SZX}, which means that blocks can have a size from 2^4 to 2^{10} bytes.

In the following examples, the notation 2:0/1/128 means that the message contains a Block2 option, NUM = 0, M = 1, and SZX = $log_2 128 - 4 = 3$.

Figure 2.18 shows the use of the Block2 option to perform the block-wise transfer of a resource requested by a client. The client starts by requesting the */status* resource on the server by issuing a GET request. The server, in a piggy-backed response, replies with a 2.05 response, which includes a Block2 option, with the values 0/1/128. This means that this is the first block (NUM = 0), another block will follow (M = 1), and that the size of the block is 128 bytes. After receiving the first block, the client performs another request for the next block. To do

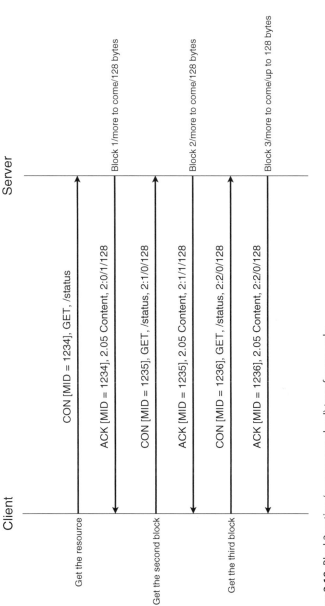

Figure 2.18 Block2 option (response payload) transfer example.

so, it includes a Block2 option in the request, specifying the parameters of the block: 1/0/128. NUM = 1 means that the requested block is the second. M does not affect the request. The client also asks for a block 128 bytes in size. The server responds in the same way as before, with a Block2 option with value 1/1/128. The client then requests the next block. In this case, the response's Block2 option has a value of 2/0/128. M = 0 indicates that this is the last block. In this case, the reported size of 128 bytes indicates that the block will be of up to 128 bytes in size. This response terminates the blockwise transfer.

Figure 2.19 shows the use of the Block1 option to perform a blockwise transfer of a resource that a client intends to post. The client starts by sending the representation of a *status* resource on the server by issuing a PUT request. The request contains a Block1 option, with the values 0/1/128. This means that this is the first block (NUM = 0), another block will follow (M = 1), and that the size of the block is 128 bytes. The server, in a piggy-backed response, replies with a 2.31 response (meaning that it is ready to accept more blocks), which includes a Block1 option, with the values 0/1/128. After sending the first block, the client performs another request to transfer the next block. To do so, it includes a Block1 option in the request, with value 1/1/128. NUM = 1 means that the requested block is the second. M does not affect the request. The client also asks for a block 128 bytes in size. The server responds in the same way as before, with a Block1 option with value 1/1/128. The client then requests the last block, including a Block1 option with value 2/0/128. M = 0 indicates that this is the last block. In this case, the reported size of 128 bytes indicates that the block will be of up to 128 bytes in size. The server sends a 2.04 response, which terminates the blockwise transfer.

Multicast communication CoAP supports multicast, which targets multiple endpoints with a single request. For example, a single POST request targeting a multicast group address, such as POST *all.floor1.building.example.com/status/lights*, can be used to switch on all lights in a building with a single request, rather than requiring a client to issue multiple unicast requests targeting each server (as shown in Figure 2.20). RFC 7390 [11] defines mechanisms to manage CoAP groups.

Resource discovery and resource directory
Resource discovery is based on discovering links (URIs) to resources hosted by a CoAP service endpoint. CoAP supports resource discovery through a standard mechanism based on web linking [12] and

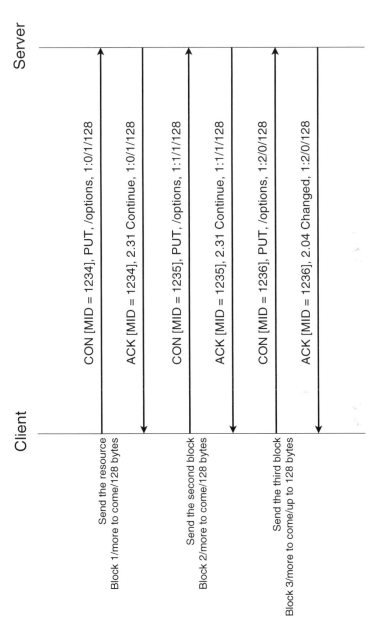

Figure 2.19 Block1 option (request payload) transfer example.

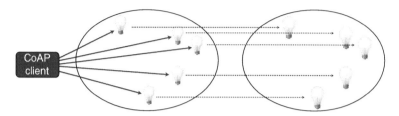

Figure 2.20 CoAP group communication.

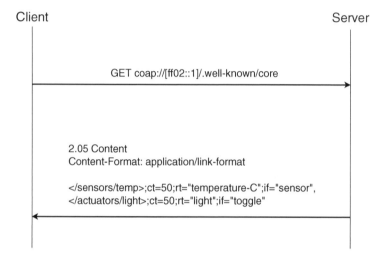

Figure 2.21 Resource discovery in CoAP.

the CoRE Link Format [13]. A client can issue a request for the /.well-known/core path on a server (or targeting a CoAP multicast group) and receive a list of available resources on the endpoint, formatted in CoRE Link Format (Figure 2.21). Each link contains a list of attributes that describe the resource being linked (e.g., *obs* means observable; *rt* indicates the resource type; *ct* indicates the content type). The *if* attribute defines the set of methods that the resource accepts (CoRE interface). More details on CoRE Link Format and CoRE interfaces can be found in Sections 3.9.1 and 3.9.2, respectively.

Sometimes direct discovery of resources through /.well-known/core is not feasible. For instance, nodes may be sleeping or multicast traffic may be inefficient. In order to simplify the task of resource discovery, a *resource directory* (RD) can be used. An RD is a CoAP endpoint

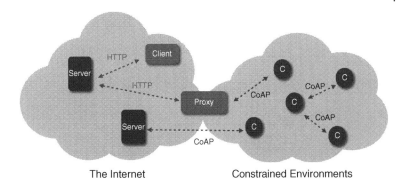

The Internet   Constrained Environments

Figure 2.22 HTTP-to-CoAP proxying.

that can be used as a registry to register, maintain, lookup and remove resource descriptions [14]. It is a centralized registry that contains CoRE links to resources hosted on several endpoints. The RD supports a function set for adding, updating, and removing links and a function set for lookup operations. The RD can be queried to filter results on the basis of specific attributes (say, a specific resource type).

HTTP/CoAP proxying

There are several reasons that motivate the use of HTTP/CoAP proxying functionalities. Figure 2.22 shows a complex hybrid Internet/IoT scenario with smart objects in constrained environments interacting with traditional hosts on the Internet. This scenario involves the use of different application protocols, such as HTTP and CoAP. Direct CoAP communication might occur, but not all traditional Internet clients may support CoAP or be aware of the fact that a particular resource resides in a constrained network. In order to allow the necessary cross-protocol communication, an intermediate proxy node must be introduced.

Other motivations for the introduction of this network element are:

- to shield the constrained network from the outside, say for security reasons such as protection against DoS attacks;
- to support integration with the existing web through legacy HTTP clients;
- to ensure high availability on resources through caching;
- to reduce network load;

- to support data formats that might not be suitable for constrained applications, such as XML.

RFC 7252 defines the following terms:

- *origin server*: a CoAP server on which a given resource resides or is to be created (see Figure 2.23a);
- *intermediary*: a CoAP endpoint that acts both as a server and as a client towards an origin server (see Figure 2.23b);
- *proxy*: an intermediary that forwards requests and relays back responses; it can perform caching and protocol translation (e.g. HTTP-to-CoAP) (see Figure 2.23c);
- *CoAP-to-CoAP proxy*: a proxy that maps a CoAP request to another one;
- *cross-proxy*: a proxy that performs protocol translation from/to CoAP to/from another protocol, such as HTTP;
- *HTTP-to-CoAP proxy*: a proxy that translates HTTP requests to CoAP requests (Figure 2.23d).

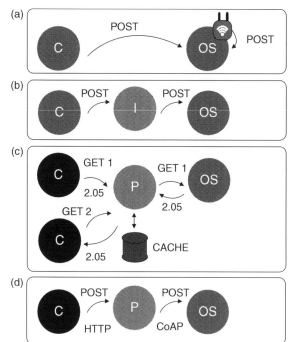

Figure 2.23 (a) Direct endpoint communication to origin server; (b) communication through intermediary node; (c) proxy-based communication. (d) HTTP-to-CoAP proxy.

A HTTP-to-CoAP proxy can thus be used to allow incoming HTTP request generated by HTTP clients to be translated into CoAP requests that can be forwarded and served by CoAP servers. A reverse translation is then required, allowing CoAP responses to be transformed into HTTP responses that can be sent back to the client. CoAP has been defined with the goal of mapping to HTTP easily, allowing easy integration with the web. CoAP-to-HTTP translation is fairly straightforward, as CoAP is a logical subset of HTTP. CoAP methods, response codes, and options have many similarities to HTTP.

RFC 8075 [15] defines rules for mapping HTTP messages to CoAP messages. These rules apply to URI mappings, request method mappings, response code mappings, and HTTP header to CoAP option mappings.

URI mapping   HTTP-to-CoAP URI mapping is based on URI encapsulation (a CoAP URI is included in an HTTP URI), as shown in Figure 2.24.

Request method mapping   Some HTTP methods and CoAP methods are easily mapped:

- HTTP GET ⇔ CoAP GET.
- HTTP POST ⇔ CoAP POST.
- HTTP PUT ⇔ CoAP PUT.
- HTTP DELETE ⇔ CoAP DELETE.

Other HTTP methods cannot be mapped.

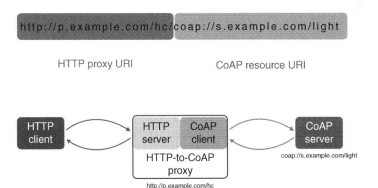

Figure 2.24 HTTP-to-CoAP URI mapping.

Table 2.3 HTTP/CoAP response code mappings.

CoAP response code	HTTP status code
2.01 Created	201 Created
2.02 Deleted	200 OK
	204 No Content
2.03 Valid	304 Not Modified
	200 OK
2.04 Changed	200 OK
	204 No Content
2.05 Content	200 OK
4.00 Bad Request	400 Bad Request
4.02 Bad Option	400 Bad Request
4.04 Not Found	404 Not Found
4.05 Method Not Allowed	405 Method Not Allowed
5.00 Internal Server Error	500 Internal Server Error
5.01 Not Implemented	501 Not Implemented
5.02 Bad Gateway	502 Bad Gateway

Response code mapping Table 2.3 shows the mapping between CoAP response codes and HTTP response codes.

Open issues The following are some questions related to protocol mapping that derive from the different natures of HTTP and CoAP:

- What HTTP features are not present in CoAP and vice versa
- How can HTTP methods that do not exist in CoAP be mapped (e.g., HEAD, OPTIONS)?
- What are resource observations in HTTP?
- What are group communications in HTTP?

2.2.5.2   CoSIP Protocol Specification

There are many applications in both constrained and non-constrained environments that feature non-request/response communication models. Some of these applications require the creation and management of a "session", a term that we use to refer to any exchange of data between an association of participants. For two participants, the session may involve the sending of one or (probably) more data packets from one participant to the other, in one or both directions. Unidirectional sessions may be initiated by either the sender or the receiver. Examples of sessions in IoT scenarios may be the data flow generated by a sensor (measurement samples) and sent to recipient for further processing, or data streams exchanged by two interacting toys.

Although in principle CoAP encapsulation could also be used for carrying data in a non-request/response fashion, for example by using the CoAP POST request in non-confirmable mode, or by using the CoAP "observation" model, it is evident that it would be much more efficient to set up a session between constrained nodes first, and then perform a more lightweight communication without carrying unnecessary CoAP header fields for each data packet. The data communication would be in accord with the network, transport, and application parameters negotiated during the session setup.

Session Initiation Protocol (SIP) is the standard application protocol for establishing application-level sessions [5]. It allows the endpoints to create, modify, and terminate any kind of (multi)media session: VoIP calls, multimedia conferences, or data communication. Once a session has been established, the media are typically transmitted using other application-layer protocols, such as RTP and RTCP [16], or as raw UDP data, directly between the endpoints, in a peer-to-peer fashion. SIP is a text protocol, similar to HTTP, which can run on top of several transport protocols, such as UDP (default), TCP, or SCTP, or on top of secure transport protocols such as TLS and DTLS. Session parameters are exchanged as SIP message payloads; a standard protocol used for this purpose is the session description protocol [17]. The SIP protocol also supports intermediate network elements, which are used to allow endpoint registration and session establishment. Examples include SIP proxy servers and registrar servers. SIP also defines the concepts of transaction, dialog, and call as groups of related messages, at different abstraction layers.

Although SIP has been defined for Internet applications, we may imagine also using it in constrained IoT scenarios. Note that SIP already includes mechanisms for subscribe/notify communication paradigms [18] and for resource directories, which are particularly useful in IoT scenarios, and for which proper CoAP extensions are currently being specified [9, 14].

The main drawback of using the standard SIP protocol in constrained environments is the large size of text-based SIP messages (compared to other binary protocols such CoAP), and the processing load required for parsing such messages.

A constrained version of SIP, named "CoSIP", designed to allow constrained devices to instantiate communication sessions in a lightweight and standardized fashion and can be adopted in M2M application scenarios, has been proposed. Session instantiation can include a negotiation phase, in which the parameters that will be used for all subsequent communication are agreed. As proposed, CoSIP is a binary protocol, which maps to SIP, just as CoAP does to HTTP. CoSIP can be adopted in various application scenarios, such as service discovery and pub/sub applications.

Related work on session initiation

Smart objects are typically required to operate using low-power and low-rate communication methods, featuring unstable (lossy) links, such as IEEE 802.15.4. These are usually termed low-power wireless personal area networks (LoWPANs) or low-power and lossy networks (LLNs). The Internet Engineering Task Force (IETF) has set up several working groups in order to address issues related to bringing IP connectivity to LoWPAN smart objects. In particular, the 6LoWPAN (IPv6 over Low-power WPAN) working group [19] was chartered to define mechanisms to optimize the adoption of IPv6 in LoWPANs and the ROLL (Routing over Low-power and Lossy Networks) working group [20] was formed to develop optimal IPv6 routing in LLNs. Finally, the CoRE (Constrained RESTful Environments) working group [21] was chartered to provide a framework for RESTful applications in constrained IP networks. It is working on the definition of a standard application-level protocol, namely CoAP, which can be used to let constrained devices communicate with any node, either on the same network or on the Internet, and provides a mapping to HTTP REST APIs. CoAP is intended to provide, among others, create-read-update-delete (CRUD) primitives for resources of constrained devices and pub/sub communication capabilities. While the work on CoAP is already at an advanced stage, the CoRE working group is also investigating mechanisms for discovery and configuration, but work on these issues is still at an early stage and therefore open to proposals.

The "observer" CoAP extension [9] allows CoAP clients to observe resources (via a subscribe/notify mechanism) and to be notified when the state of the observed resource changes. This approach requires the introduction of a new CoAP *Observe* option to be used in GET requests in order to let the client register its interest in the resource. The server will then send "unsolicited" responses back to the client, echoing the token it specified in the GET request and reporting an *Observe* option with a sequence number used for reordering purposes. As we will describe later, we envision that the instantiation of a session could significantly reduce the number of transmitted bytes, since, after the session has been established, only the payloads need be sent to the observer, thus eliminating the overhead due to the inclusion of the CoAP headers in each notification message.

As for service discovery, the CoRE working group has defined a mechanism, called a resource directory (RD) [14], to be adopted in M2M applications. An RD is necessary because of the impracticality of direct resource discovery, due to the presence of duty-cycled nodes and unstable links in LLNs.

Each CoAP server must expose an interface /.well-known/ core to which a client can send requests for discovering available resources. The CoAP server will reply with the list of resources and, for each resource, an attribute that specifies the format of the data associated with it. The CoAP protocol, however, does not specify how a node joining the network for the first time must behave in order to announce itself to the RD node. In RFC 7390 [11], this functionality is extended to multicast communications. In particular, multicast resource discovery is useful when a client needs to locate a resource within a limited, local scope, and that scope supports IP multicast. A GET request to the multicast address specified by the standard is made for /.well-known/core. Of course multicast resource discovery works only within an IP multicast domain and does not scale to larger networks that do not support end-to-end multicast.

The registration of a resource in the RD is performed by sending a POST request to it. Discovery can be accomplished by issuing a GET request to the RD, targeting the .well-known/core URI. This discovery mechanism is totally self-contained in CoAP as it uses only CoAP messages.

The CoSIP protocol provides an alternative mechanism to register resources on an RD, which may also be called a CoSIP registrar server. The advantage of using a CoSIP-based registration mechanism is that it might be possible to register resources other than those reachable through CoAP, thus providing a scalable and generic mechanism for service discovery in constrained applications with a higher degree of expressiveness, such as setting an expiration time for the registration.

CoSIP

In both constrained and non-constrained environments, there are many applications in which it may either be necessary or simply advantageous to negotiate an end-to-end data session. In this case the communication model consists of a first phase in which one endpoint requests the establishment of a data communication and, optionally, both endpoints negotiate communication parameters (transfer protocols, data formats, endpoint IP addresses and ports, encryption algorithms and keying materials, and other application specific parameters) of the subsequent data sessions. This may be useful for both client-server and peer-to-peer applications, regardless of whether the data sessions evolve according to a request/response model. The main advantage is that all such parameters, including possible resource addressing, may be exchanged in advance, while no such control information is required during data transfer. The longer the data sessions, the more advantageous this approach is compared to per-message control information. In addition, for data sessions that may vary in format or other parameters over time, such changes may be supported by performing session renegotiation.

A standard way to achieve all this in an IP-based network is by using SIP [5]. SIP has been defined as a standard protocol for initiating, modifying and tearing down any type of end-to-end multimedia session. It is independent of the protocol used for data transfer and from the protocol used for negotiating the data transfer (such a negotiation protocol can be encapsulated transparently within the SIP exchange). In order to simplify the implementation, SIP reuses the message format and protocol fields of HTTP. However, in contrast to HTTP, SIP works by default on UDP, by directly implementing all mechanisms for a reliable transaction-based message transfer. This is an advantage in duty-cycled constrained environments, where problems may arise when trying to use connection-oriented transports, such as TCP. However, SIP may also run on other transport protocols, such as TCP, SCTP, TLS, or DTLS.

Unfortunately, SIP derives from HTTP the text-based protocol syntax that, even if it simplifies the implementation and debugging, results in larger message sizes and bigger processing costs and probably with larger source code sizes (RAM footprint) required for message parsing. Note that the SIP standard also defines a mechanism for reducing the overall size of SIP messages; this is achieved by using a compact form of some common header field names. However, although this allows for a partial reduction of the message size, it may still result in big messages, especially when compared to other binary formats, for example those defined for CoAP.

For this reason we have tried to define and implement a new binary format for SIP in order to take advantages of the functionalities already defined and supported by SIP methods and functions, together with a very compact message encoding. We naturally called such new protocol CoSIP, standing for Constrained Session Initiation Protocol, or simply Constrained SIP. Due to the protocol similarities between SIP and HTTP, in order to maximize the reuse of protocol definitions and source code implementations, we decided to base CoSIP on the same message format as defined for CoAP, thanks to the role that CoAP plays with respect to HTTP. However, it is important to note that, while CoAP must define new message exchanges, mainly due to the fact that it has to operate in constrained and unreliable network scenarios over the UDP transport protocol, and while HTTP works over TCP, CoSIP completely reuses all of the SIP message exchanges and transactions already defined by the SIP standard, since SIP already works over unreliable transport protocols such as UDP.

SIP is structured as a layered protocol. At the top there is the concept of dialog: a peer-to-peer relationship between two SIP nodes that persists for some time and facilitates sequencing of different request–response exchanges (transactions). In CoAP there is no concept equivalent to SIP dialogs, and, if needed, it has to be explicitly implemented at application level. Under the dialog there is the transaction layer: the message exchange that comprises a client request, the ensuing optional server provisional responses and the server's final response. The concept of a transaction is also present in CoAP: requests and responses are bound and matched through a token present in the message header field. Under the transaction there is the messaging layer where messages are effectively formatted and sent through an underlying non-SIP transport protocol (such as UDP or TCP).

Instead of completely re-designing a session initiation protocol for constrained environments, we propose to reuse SIP's layered architecture, by simply re-defining the messaging layer in a constrained-oriented binary encoding. To this end, we propose to

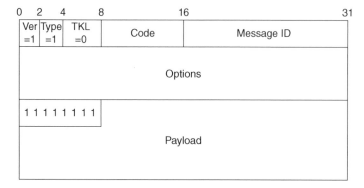

Figure 2.25 CoSIP message format.

reuse the same CoAP message syntax [7]. Figure 2.25 shows the CoSIP message format derived from CoAP. A CoSIP message contains, in sequence:

- the 2-bit version field (set to 1, i.e. CoSIP version 1);
- the 2-bit type field (set to 1 = Non-confirmable);
- the 4-bit CoAP TKL field (set to 0);
- the 8-bit Code field that encodes request methods (for request messages) and response codes (for response messages);
- the 16-bit CoAP message ID field;

possibly followed by by more option fields. If a CoSIP message body is present, as in CoAP it is appended after the options field, prefixed by an 1-byte marker (0xFF) that separates CoSIP header and payload. Options are encoded, as in CoAP, in Type-Length-Value (TLV) format and encode all CoSIP header fields (From, Via, Call-ID, etc.) included in the CoSIP message.

Since CoSIP re-uses the transaction layer of SIP, no CoAP optional Token field is needed [7] and the TKL (Token length) field can be permanently set to 0. Moreover, since CoSIP already has reliable message transmission (within the transaction layer), no Confirmable (0), Acknowledgement (2), or Reset (3) message types are needed, and the only type of message that must be supported is Non-confirmable (1).

A comparison of the layered architectures of CoSIP and SIP is shown in Figure 2.26.

Besides the above binary message, a CoSIP message can be virtually seen as a standard SIP message, formed by one request-line or one

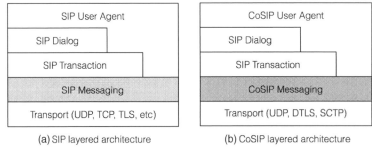

(a) SIP layered architecture   (b) CoSIP layered architecture

Figure 2.26 Comparison of the layered architectures of: (a) SIP and (b) CoSIP.

status-line (depending if the message is a request or a response), followed by a sequence of SIP header fields, followed by a message body, if preset. In particular, SIP header fields are logically the same as in the standard SIP protocol, but encoded in the corresponding CoSIP Options. For each SIP header field, a different option number has been set, and a suitable encoding mechanism has been defined. In particular, general rules that we followed are:

- IP addresses are encoded as a sequence of 5 bytes for IPv4 and 17 bytes for IPv6, where the first byte discriminates the type of address, i.e. 1 = IPv4 address, 2 = IPv6 address, 3 = FQDN (fully qualified domain name).
- For header field parameters, when possible, the parameter name is implicitly identified by the position of its value in the corresponding binary-encoded CoSIP option; otherwise, parameter names are substituted by parameter codes. In the latter case the parameter is encoded as type-value pair (for fixed size values) or type-length-vale tuples (for variable size values).
- Random tokens, such as SIP "branch" values, SIP "from" and "to" tags, "call-id", etc. are generated as arrays of a maximum 6 bytes.

One problem in reusing the current CoAP message format [7] is that in CoAP the 8-bit code field is used to encode all possible request methods and response codes. In particular, in CoAP, for response messages, the 8-bit code field is divided into two subfields:

- the first three bits (class) encodes the CoAP response classes 2xx (Success), 4xx (Client error), and 5 (Server error);

- the remaining 5 bits (details) encode the sub-type of the response within a given class type. For example a 403 "Forbidden" response is encoded as 4 (class) and 03 (details).

Unfortunately, this method limits the number of possible response codes that can be used (for example, using only 5 bits for the details subfield does not allow the direct encoding of response codes such as 480 "Temporarily unavailable" or 488 "Not acceptable here"). In CoSIP, we overcome this problem by encoding within the code field only the response class (2xx, 4xx, etc.) and by adding an explicit option field, called response-code, which encodes the complete response code (e.g. 488), including the response sub-type (88, in the case of response code 488). The size of the response-code option is 2 bytes. Moreover, in order to support all SIP/CoSIP response codes we also added the classes 1xx (Provisional) and 3xx (Redirect) used in SIP.

IoT application scenarios

In this section, we will describe the most significant IoT applications, in order to provide an overview of the capabilities and typical usage of the CoSIP protocol. In all the scenarios, we consider a network element called an IoT gateway, which includes also a HTTP/CoAP proxy that can be used by nodes residing outside the constrained network to access CoAP services.

CoAP service discovery CoSIP allows smart objects to register the services they provide to populate a CoSIP registrar server, which serves as an RD. The term registrar server is interchangeable with RD here.

Figure 2.27 shows a complete service registration and discovery scenario enabled by CoSIP. We consider a smart object that includes a CoAP server, which provides one or more RESTful services, and a CoSIP agent, which is used to interact with the CoSIP registrar server. The smart object issues a REGISTER request (a), which includes registration parameters, such as the address of record (AoR) of the CoAP service and the actual URL that can be used to access the resource (contact address). Note that, while the original SIP specification states that the To header *must* report a SIP or SIPS URI, CoSIP allows any scheme URI to be specified in the To header, for example a CoAP URI. Upon receiving the registration request, the registrar server stores the AoR-to-contact address mapping in a location database and then sends a 200 OK response.

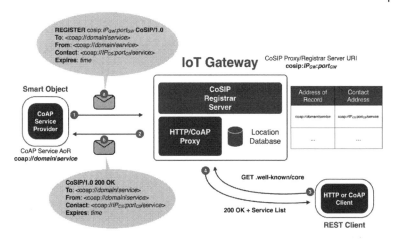

Figure 2.27 CoAP service discovery. The numbers indicate the order of exchanged messages.

When a REST client, either CoAP or HTTP, is willing to discover the services, it can issue a GET request targeting the `.well-known/core` URI, which is used as a default entry point to retrieve the resources hosted by the RD, as defined in RFC 6690 [13]. The GET request is sent to the HTTP/CoAP proxy, which returns a 200 OK response (in the case of HTTP) or a 2.05 Content response containing the list of services in the payload (in the case of CoAP).

Session establishment A session is established when two endpoints need to exchange data. CoSIP allows the establishment of session in a standard way without binding the session establishment method to a specific session protocol. For instance, CoSIP can be used to negotiate and instantiate a RTP session between constrained nodes. Once a session has been established, the data exchange between the endpoints occurs (logically) in a peer-to-peer fashion.

Figure 2.28 shows how CoSIP can be used to establish a session between two endpoints. Let us assume an IoT agent (IoT-A_1) identified by the CoSIP URI cosip:user1@domain, and which includes at least a CoSIP agent, has registered its contact address to an IoT gateway in the same way as described in the previous subsection, on CoAP service discovery (steps 1 and 2). If another IoT-A_2 cosip:user2@domain wants to establish a session with IoT-A_1, it will send a suitable INVITE request to the IoT gateway, which will act as

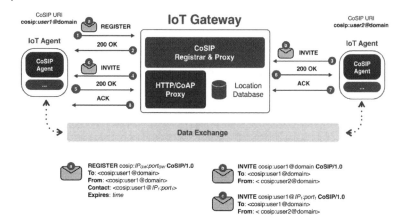

Figure 2.28 CoSIP session establishment.

a CoSIP proxy, relaying the request to IoT-A$_1$ (steps 3 and 4). IoT-A$_1$ will then send a 200 OK response to IoT-A$_2$ (steps 5 and 6), which will finalize the session creation by sending an ACK message to IoT-A$_2$ (steps 7 and 8).

At this point the session has been set up and data flow between IoT-A$_1$ and IoT-A$_2$ can occur directly. The session establishment process can be used to negotiate communication parameters, for instance by encapsulating Session Description Protocol (SDP) [17] or an equivalent in the message payload. As we will show in the protocol evaluation section below, setting up a session, rather than using CoAP, both in a request/response or subscribe/notify paradigm, is a very efficient way to avoid the overhead due to carrying headers in each exchanged message, since eventually only the payloads will be relevant for the application.

Subscribe/notify applications IoT scenarios typically involve smart objects, which may well be battery-powered. It is crucial to adopt energy-efficient paradigms for OS tasks, application processing, and communication. In order to minimize the power consumed, duty-cycled smart objects are used. Sleepy nodes, especially those operating in LLNs, are not guaranteed to be reachable, so it is more appropriate for smart objects to use a subscribe/notify (pub/sub) approach to send notifications regarding the state of their resources, rather than receiving and serving incoming requests. Such behavior

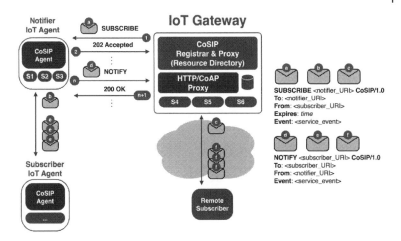

Figure 2.29 Subscribe/notify applications with CoSIP.

can be achieved by leveraging on the inherent capabilities of SIP, and therefore of CoSIP, as sketched in Figure 2.29.

The depicted scenarios consider several pub/sub interactions: notifications can be sent either by a notifier IoT agent (IoT-A_N) or by an IoT gateway, and subscribers can be either subscriber IoT agents (IoT-A_S), IoT gateways, or generic remote subscribers. Let us assume that all the notifiers have previously registered with their CoSIP registrar server (this step is also called the publishing phase in a typical pub/sub scenario). The standard subscription/notification procedure is the following:

1) The subscriber sends a SUBSCRIBE request to the notifier, also specifying the service events it is interested in.
2) The notifier stores the subscriber's URI and event information and sends a 200 OK response to the subscriber.
3) Whenever the notifier's state changes, it sends a NOTIFY request to the subscriber.
4) The subscriber sends a 200 OK response back to the notifier.

Figure 2.29 reports all the use cases when pub/sub might be used. An IoT-A_S can subscribe to the service of an IoT-A_N in the same network, if it is willing to perform some task, such as data/service aggregation. The IoT gateway can subscribe to an IoT-A_N in order to collect sensed data, say to store them in the cloud, without the need to periodically poll for data. Finally, the IoT gateway itself might be a notifier for remote subscribers that are interested in notifications for specific services provided by the gateway, which may or may not be the same of existing IoT-A_N nodes managed by the gateway. Note that, it might be possible to have interactions with legacy SIP agents if the IoT gateway is also able to perform SIP/CoSIP proxying.

The adoption of CoSIP in IoT scenarios makes it easy to set up efficient pub/sub-based applications in a standard way, thus allowing for seamless integration and interaction with the Internet. Moreover, the valuable experience gained in recent years with SIP, both in terms of technologies and implementation, can be reused to speed up the implementation and deployment of session-based applications.

Protocol evaluation

In order to evaluate the performance of CoSIP, an implementation of the protocol has been developed together with some test applications. We decided to focus on network performance as a metric by measuring the amount of network traffic generated by the test applications. The CoSIP protocol was implemented in Java, because of its simplicity, cross-platform support, and the existence and availability of SIP and CoAP libraries [22–24]. The source code of the CoSIP implementation is freely available [25–27].

The results show that there are many advantages to using CoSIP, both in constrained and non-constrained applications. The first evaluation compares CoSIP and SIP in terms of bytes transmitted for the signaling related to the instantiation and termination of a session. Each CoSIP request and response message is separately compared with its SIP counterpart. The results are illustrated in Figure 2.30. Table 2.4 shows the compression ratio for each CoSIP/SIP message pair. Regarding the session as a whole, CoSIP yields an overall compression ratio of slightly more than 0.55.

Another evaluation showed the advantage of using sessions in constrained applications. Figure 2.31 shows the amount of network traffic (in bytes) generated by two constrained applications: the first application uses CoSIP to establish a session and then performs the data exchange by sending the payloads over UDP; the second is a standard CoAP-based application where the communication occurs between a CoAP client and a CoAP server, using confirmed CoAP POST requests. In both cases data is sent at the same rate of one data message every 2s. The figure shows that the lightweight CoSIP session

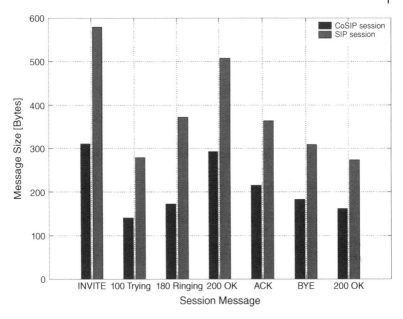

Figure 2.30 Transmitted bytes for CoSIP and SIP session (signaling only).

Table 2.4 Comparison between CoSIP and SIP signaling (bytes per message) for session instantiation and establishment.

Message type	CoSIP (bytes)	SIP (bytes)	Compression Ratio
INVITE	311	579	0.537
100 Trying	141	279	0.505
180 Ringing	173	372	0.465
200 OK	293	508	0.577
ACK	216	363	0.595
BYE	183	309	0.592
200 OK	162	274	0.591

is instantiated in a very short period of time and, after the session has been established, few bytes are exchanged between the endpoints. On the other hand, the CoAP-based application has no overhead at the beginning due to the instantiation of the session but, soon afterwards, the amount of traffic generated by the application exceeds that of the

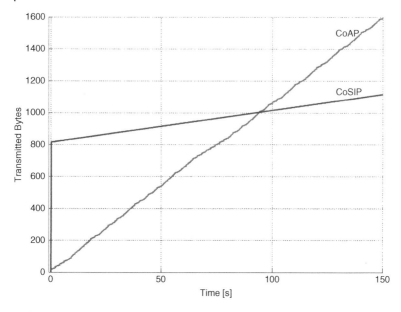

Figure 2.31 Transmitted bytes in a CoSIP session vs. CoAP confirmed POST requests and responses.

CoSIP-based application, since in the CoAP-based scenario, data is exchanged within CoAP messages, resulting in an unnecessary CoAP overhead.

Note that in the depicted scenario the CoSIP signaling used for session initiation includes all SIP header fields normally used in standard non-constrained SIP applications; that is, no reduction in term of header fields has been performed. Instead, for the CoAP application, we considered only mandatory CoAP header fields, resulting in the best-case scenario for CoAP in terms of CoAP overhead (minimum overhead). This means that in other CoAP applications, the slope of the line could become even steeper, thus reducing the time when the break-even point with CoSIP is reached.

Conclusions

Here we have introduced a low-power protocol called CoSIP, for establishing sessions between two or more endpoints targeting constrained environments. Many applications, both in constrained and non-constrained scenarios, do benefit from establishing a session between the participants in order to minimize the communication overhead and to negotiate parameters related to the data exchange that will occur. The CoSIP protocol is a constrained version of the SIP protocol, designed to minimize the amount of network traffic and therefore energy consumption, and targeted at IoT scenarios.

A similar effort in trying to minimize the amount of data in IoT and M2M applications is being carried on in standardization organizations, such as the IETF CoRE working group, which is currently defining a protocol (CoAP) to be used as a generic web protocol for RESTful constrained environments, and which maps to HTTP. Similarly, in this work we have applied the same approach to define a protocol for session instantiation, negotiation, and termination. We have described some interesting IoT scenarios that might benefit from using such a protocol, namely service discovery, session establishment, and services based on a subscribe/notify paradigm. A Java-language implementation of CoSIP has been developed and tested to evaluate the performance of the new protocol, by measuring the number of transmitted bytes compared to solutions based on SIP and CoAP. The results show that applications that use CoSIP can outperform other SIP- and CoAP-based applications in terms of generated network traffic: SIP signaling can be compressed by nearly 50% using CoSIP, and long-running applications that may use CoAP for sending the same type of data to a given receiver may be better implemented with CoSIP, since no CoAP overhead has to be transmitted along with each transmitted data message, leading to a packet size and per-packet processing reduction; packet size reduction in turn may reduce the need for packet fragmentation (in 6LoWPAN networks) and the energy consumption of the nodes involved in the data exchange.

Future work will include exhaustive experimentation, both in simulation environments and a real-world testbeds comprising a variety of heterogeneous devices. These are currently being set up at the Department of Information Engineering of the University of Parma. The aim is to evaluate the performance of the CoSIP protocol both in terms of energy consumption and delay. The tests will focus on the time required to set up a session in different scenarios, such as in IEEE 802.15.4 multi-hop environments, and the measurement of energy consumption compared to standard CoAP communication.

Two different perspectives will be analyzed: end-to-end delay between the actual session participants and energy consumption on the intermediate nodes which will be indirectly involved in the session, responsible for multi-hop routing at lower layers. The target platforms will be both constrained and non-constrained devices for session participants and relay nodes, in order to provide a thorough evaluation in heterogeneous devices operating under different conditions.

2.3   The Industrial IoT

IIoT still lacks a reference networking/communication platform. Several initiatives are being developed: in the following, we comment on a few relevant ones.

The German Plattform Industrie 4.0 is a candidate to become a European standard platform. This process is taking place within EU institutions, and individual European countries have their own industrial transformation projects in which the IIoT takes center stage, including:

- Smart Factory (The Netherlands)
- Factory 4.0 (Italy)
- Industry of the Future (France).

Other major efforts include the Japanese Robot Revolution initiative and the Industrial Internet Consortium (IIC). The latter is a consortium co-founded by US industrial giant GE, which also coined the term Industrial Internet and is one of the major players in the IIoT. The IIC today is busy mainly with the promotion of the IIoT, in which data is used in order to improve operations, enhance service and detect new opportunities. The IIC collaborates with the Industry 4.0 Platform.

Just like the Industry 4.0 Platform, the Internet of Things Consortium has developed a framework called the Industrial Internet Reference Architecture (IIRA). The first version was released in 2015 and version 1.8 of the IIRA was published in January 2017. It aims to help all sorts of experts who are involved in IIoT projects to consistently design IIoT solution architectures and deploy interoperable IIoT systems. On top of the IIRA model, in February 2017 the IIC also published the Industrial Internet Connectivity Framework (IICF).

Other initiatives are also being considered. For instance, the Organization for Machine Automation and Control, the OPC Foundation, and PLCopen, which have worked independently on different aspects of automation standardization, are now combining efforts to create companion specifications for the standards and protocols they have already developed in order to allow seamless IIoT interoperability. For example, the OPC Foundation's "Unified Architecture" (OPC UA) is an industrial interoperability framework. It delivers information modeling with integrated security, access rights, and all communication layers to provide plug and play machine-to-machine (M2M) communication inside factories. It is scalable across the plant floor and from sensor to IT enterprise and cloud scenarios. OPC and PLCopen – which is focused around IEC 61131-3, the only global standard for industrial control programming – worked together to define a set of function blocks to map the IEC 61131-3 standard for industrial controls programming to the OPC UA information communication model.

3

Interoperability

3.1 Applications in the IoT

Costs, limited size and minimal energy consumption are a few of the reasons that IoT devices have limited computational capabilities. Because of these functional and economic requirements, smart objects, especially those that are battery-powered, cannot afford to have heavy processing loads and use expensive communication protocols. On the one hand, limited processing capabilities means that it is hard to process large messages. On the other hand, less processing means lower energy consumption. As a result, IoT devices typically need to minimize the amount of transmitted data.

Devices that operate in low-power and lossy networks (LLNs), can greatly benefit from lightweight protocols. Large messages result in more fragments (6LoWPAN), which introduce overhead: due to the unstable nature of LLNs, transmitting more fragments can require multiple retransmissions before a whole message can be successfully reconstructed by the receiver. These retransmissions may result in more delay and energy consumption.

Communication protocols are specific communication paradigms, which can be classified into two categories: request/response and publish/subscribe (pub/sub). Specific application scenarios have requirements that drive the choice of the most suitable communication paradigm (and protocol). The question of which architecture fits best does not have a clear "one-for-all" answer.

Internet of Things: Architectures, Protocols and Standards, First Edition.
Simone Cirani, Gianluigi Ferrari, Marco Picone, and Luca Veltri.
© 2019 John Wiley & Sons Ltd. Published 2019 by John Wiley & Sons Ltd.

3.2   The Verticals: Cloud-based Solutions

The early days of the IoT were characterized by the adoption of a very simplistic approach to interconnecting devices: by relying on the availability of cloud services, all makers needed to do was to connect things to the Internet (either through cellular networks or in many cases through an Internet-connected gateway) and send all data uplink to the cloud. The cloud service would then provide a storage facility for all data sent by devices on one side, and a HTTP-based interface for access by clients (through browser or vendor-specific mobile apps) on the other. All major cloud service providers (as illustrated in Figure 3.1), such as Amazon and Microsoft, have now entered this market and released their own cloud IoT platforms. Amazon's AWS IoT and Microsoft's Azure IoT suite are probably the most popular cloud IoT platforms. These cloud IoT platforms are an easy way for makers to deploy their applications without requiring them to invest development resources to realize a backend.

Although clearly easy to implement and very cost-effective, this approach has created the misunderstanding that the IoT can simply be built by connecting things to the Internet. This is a prerequisite for the IoT but it is not enough to actually create a worldwide network of interconnected devices. This first generation of hardware and software involved has introduced several issues because no attention

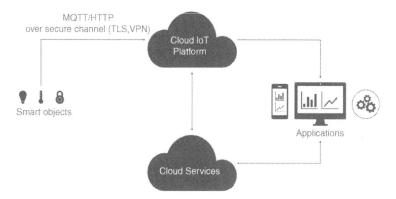

Figure 3.1 Cloud IoT platform architecture.

was paid to developing a long-term design that would actually control the network nor take into account:

- *Scalability*: The number of IoT devices is expected to reach 50 billion by 2020. At present, with the order of hundreds of millions of things, everything works, but are networks and services ready to handle the traffic generated by billions of things?
- *Availability*: What happens if an Internet connection becomes temporarily or permanently unavailable? Relying on the cloud would just make the service unavailable.
- *Interoperability*: All device-to-cloud applications do not allow direct interactions among things made by different manufacturers. Interoperability can occur only at the cloud level, through system integration of the data itself, if such data are made available to external applications.
- *Security*: Even though secure and authorized access to cloud services can be implemented in traditional ways, attackers could exploit a breach in the cloud to access a huge amount of private data or implement a DoS attack to prevent users from accessing their data.
- *Evolution of systems*: Device-to-cloud applications typically need to hard code information into things (which act as clients), thus making them less robust to changes on the server side. Any update or upgrade in the server functionality might have a destructive effect on the operations of things, which may then require a software/firmware upgrade to change how they operate (say, the endpoint they target or the data format they use).

Cloud-based solutions are never going to disappear, at least anytime soon. Nonetheless, this approach just cannot be the reference architecture for a scalable and evolutionary IoT.

3.3   REST Architectures: The Web of Things

One promising approach that is being brought to the IoT is the idea that it should be built in a similar way to the Internet. There are several reasons to use a web-based approach in the IoT. The web has been around for decades and lots of experience has been gained.

Since its public release in 1991, the World Wide Web has dramatically evolved and has become an infrastructure upon which to store documents, resources, and to build distributed applications. The most important aspects of the introduction of the web were the referencing of resources through uniform resource identifiers (URIs) [28] and the introduction of the Hypertext Transfer Protocol (HTTP) [2] as the application-layer protocol for hypermedia systems. Along with these two major pillars, other essential standards and technologies have been developed, such as the Hypertext Markup Language (HTML) for web documents, web browsers, and web servers. As the web became more and more popular, browsers integrated dynamic behavior through Javascript and Cascading Stylesheets (CSS). After all these years, HTTP is by far the most common application-layer protocol and software libraries implementing web protocols are available (web servers, HTTP clients and so on) for any programming language. Ways of building web applications are widely known and used: adopting similar approaches for the IoT could therefore take advantage of the expertise of existing developers. Moreover, the web has proved to scale extremely well: this is extremely important for the IoT, where billions of connected devices are expected to operate. The IoT can greatly benefit from all the experience gained in the development of the web and thus the use of a similar architecture would seem to be a wise design choice.

3.3.1 REST: The Web as a Platform

The web was born to be an easy to use, distributed, and loosely coupled (see below) system for sharing documents. The architecture of the web is simple enough to make it easy to build applications and manage content. The web is based on a small set of principles, yet it has proved to scale and evolve wonderfully. Thanks to these principles, the web has evolved to become a platform for building distributed systems using HTTP.

REpresentational State Transfer (REST) is the architectural style behind the web. Defined in 2000 in Roy Fielding's PhD thesis [29], REST defines a set of rules and principles that all the elements of the architecture must conform to in order to build web applications that scale well, in terms of:

- scalability (number of interacting clients)
- robustness (long-term evolution of systems).

Loose coupling means that the endpoints should contain as little information about each other as needed to work. All necessary missing information should be collected while interacting. The client must know very few things a-priori. The server will drive the client and pass in the information required to progress and to perform the intended operations. The more a client knows about the server, the more closely it depends on the server implementation. This is a weakness for an application because any change on the server must be matched by a change in the client, which would otherwise just break. In a highly dynamic, evolving, and gigantic environment such as the IoT, design principles that lead to create robust applications must be adopted.

3.3.1.1 Resource-oriented Architectures

REST is based on the concept of a resource. A resource can be defined as any relevant entity in an application's domain that is exposed on the network. A webpage, a video, and an order on an e-commerce website can all be considered web resources. A resource is anything with which a user interacts while progressing toward some goal. Anything can be mapped as a resource, as long as it is meaningful for the application. Resources are characterized by some data, such as the title of the page or the items in an order.

An alternative to a resource-oriented architecture (ROA) is a service-oriented architecture (SOA). SOAs have been around for many years and have become a reference for many legacy business-oriented systems. A SOA refers to an architecture where two endpoints communicate through a pre-defined set of messaging contracts. A client starts interacting with a server by retrieving the list of available services and how these can be mapped to HTTP messages, in a Web Service Definition Language (WSDL) document. In essence, the WSDL maps a message to a method call on the server. Remote method calls are contained in a SOAP (an XML specification) included in the body of messages. The presence of a WSDL document is needed to add semantics to messages. However, this is a weakness: if a server changes its services, a client needs to get access to the new WSDL or its functionalities are invalidated. In a ROA, on the other hand, there is no endpoint exposing services; there are only resources that can be manipulated. This is critical for the robustness of the client application.

3.3.1.2 REST Architectures

The principle of separation of concerns is a fundamental of the REST architecture. According to this principle, each element of a system is responsible for a specific concern. Well-separated concerns allow for modularity, reusability, and independent evolution of the system's elements. A REST architecture builds on:

- *clients* (or user agents, such as browsers), which are the application interface and initiate the interactions
- *servers* (origin servers) host resources and serve client requests.

Intermediaries act as clients and servers at the same time. Forward proxies (known to clients) are "exit points" for a request. Reverse proxies appear as origin servers to a client, but actually relay requests.

A REST architecture is characterized by uniform interfaces: all connectors within the system must conform to this interface's constraints. Collectively, REST defines the following principles:

- identification of resources
- manipulation of resources through representations
- self-descriptive messages
- hypermedia as the engine of application states.

An application that follows the above principles is termed RESTful.

3.3.1.3 Representation of Resources

Resources are never exchanged directly by endpoints. Instead, representations of resources are exchanged between endpoints. A representation is a view of the state of the resource at a given time. This view can be encoded in one or more transferable formats, such as XHTML, Atom, XML, JSON, plain text, comma-separated values, MP3, or JPEG. Typically, the type of representation is specified in one header of the message containing the resource; for example, HTTP defines the *Content-Type* header. For the sake of compactness, from now on we will refer to the representation of a resource simply as a resource. Resources are exchanged back and forth between clients and servers. In order to be exchanged, resources must be serialized/deserialized properly at each endpoint, as shown in Figure 3.2.

The same resource can have many different representations (1:N relationship): the state of the same sensor can be described using JSON, XML, or any other suitable format.

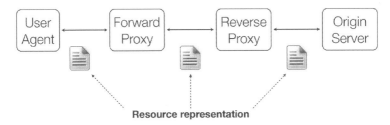

Figure 3.2 Representations of resources are exchanged between endpoints.

3.3.1.4 Resource Identifiers

In order to ensure that an application is handling the correct resource, a mechanism to identify a resource univocally in the network is necessary. uniform resource identifiers (URIs), defined in RFC 3986 [28], serve this specific need.

A URI identifies a resource univocally. A URI can be used to address a resource, so that it can be located, retrieved, and manipulated. There is a 1:N relationship between a resource and URIs: a resource can be mapped to multiple URIs, but a URI points exactly to one resource. URIs can be of two kinds:

- a uniform resource name (URN) specifies the name of a resource (e.g., `urn:ietf:rfc:2616`);
- a uniform resource locator (URL) specifies how to locate the resource, (e.g., `http://example.com/books/123`).

All URIs take the following form: `scheme:scheme-specific-part`. The scheme part defines how the rest of the URI is to be interpreted – it typically serves as an indication of the communication protocol that should be used to target the resource. For instance, URNs use the urn scheme, while web resources use the http scheme. URLs include all the information needed to successfully address the resource. A URL has the form shown in Figure 3.3.

The optional [username:password] part specifies credentials to use for authenticated access to the resource. The host and optional port include networking information needed to reach to the resource. The

http://	example.com	:8080	/people	?id=1	#address
scheme	host	port	path	query	fragment

Figure 3.3 Generic URL structure.

host can be be either an IP address or a fully qualified domain name, which must be resolved using the DNS system. The path provides information to locate the resource inside the host. The query contains matching information to filter out the result. Finally, the fragment can be used to identify a specific portion of the resource. URIs should be opaque and not expose any specific notion of the format used to represent the targeted resource. For example, `http://example.com/people/123` is a good URI, while `http://example.com/people/123.xml` and `http://example.com/people/123.json` are not.

3.3.1.5 Statelessness

An important principle of REST is statelessness. Statelessness implies that no state information must be kept on the client and server sides, thus avoiding the need to use cookies or to introduce the concept of sessions, which demand a stricter coupling between the endpoints. All requests must therefore be stateless. In order to preserve statelessness, each message must be self-descriptive. This means that all requests must contain all the information to understand the request so that servers can process them without context (about the state of the client). There is no state information maintained between clients and servers: the state of the client is contained in the request, so the server is relieved from the burden of keeping track of the client state.

3.3.1.6 Applications as Finite-state Machines

RESTful applications make forward progress by transitioning from one state to another, just like a finite-state machine (FSM), as shown in Figure 3.4.

Due to the loose coupling between clients and servers, the states of the FSM and the transitions between them are not known in advance. A state is reached when the server transfers a representation of the resource as a consequence of a client request. The next possible transitions are discovered when the application reaches a new state (gradual reveal). Resource representations that embed links are called hypermedia. These links represent the possible transitions to the next states. In essence, the state of the a resource identified by a URI is contained in the data section of the resource representation and the transition to the next states are contained in the links.

3.3.1.7 Hypermedia as the Engine of Application State

The final principle of REST is "hypermedia as the engine of application state", or HATEOAS for short. HATEOAS states that resource

Figure 3.4 Applications as finite-state machines.

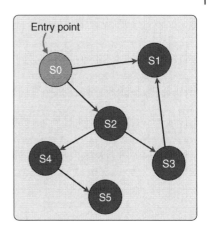

representation should include all the information to drive the client to perform the next requests to progress in the application. By doing so, a client just needs to follow the instructions that the server transmits in order to reach its goal. This guarantees that, should the server change its implementation and introduce new functionality (states and links in the FSM), the client would be unaffected by these changes and could continue to operate.

In summary, RESTful applications progress according to the following steps:

1) The client starts from an entry URI or a bookmark.
2) The response to a GET request includes a hypermedia representation.
3) The representation contains links that define the possible transitions to the next states of the FSM.
4) The client selects a link to follow and issues the next request; that is, it triggers a transition to the next state.
5) The client can also go back.

An important contribution of REST is the fact that it allows the web to be modeled as an FSM. The web is a globally distributed application, with web browsers as the clients and millions of servers that can serve requests. In the web, resources are documents (such as HTML pages). HTML includes links that can point to other documents (inside <a> tags) and is therefore a hypermedia format. The application is entered through a first URI (entered in the address bar). By clicking on a link, a new state is reached (the new document).

3.3.2 Richardson Maturity Model

The Richardson maturity model (Figure 3.5) is a classification system for web-based applications. The classification is based on levels, which determine the degree of compliance with the principles of REST. It can be used to answer the question "How RESTful is a web application?" According to the Richardson maturity model, the higher the level, the more RESTful an application is: the higher the level, the less coupling exists between clients and servers. We always have to remind ourselves that coupling between endpoints should be avoided as much as possible in order to support independent evolution of systems.

3.3.2.1 Level 0: the Swamp of POX

The first attempt to create remote procedure calls (RPCs) on the web implied the use of HTTP as a transport system for remote interactions, but without using any of the mechanisms of the web. HTTP was merely used as a tunneling mechanism for custom remote interaction. Using HTTP has great benefits, such as the use of TCP ports 80 and 443, which are typically considered safe by firewalls and therefore are not blocked. The idea behind using HTTP as a transport is to expose a service at some URI and then use a single HTTP method – namely POST – to send requests, embedding in the request payload an XML-formatted document describing the operation. In essence, clients post RPCs that trigger actions. The XML documents

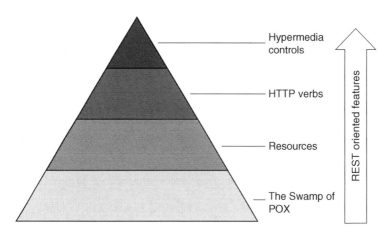

Figure 3.5 The Richardson maturity model.

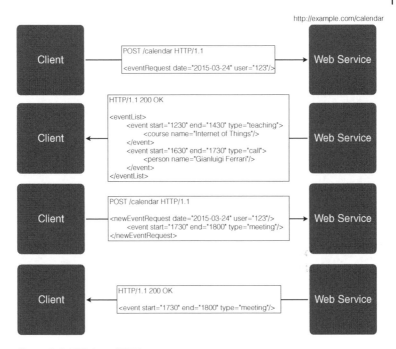

Figure 3.6 XML-based RPC.

embedded in the request and response messages describe the action and the result of that action, respectively.

All the semantics of an interaction are strictly tied to the syntax that clients and servers use. This is why Level 0 applications are in "the Swamp of POX" (plain old XML). This is depicted in Figure 3.6.

The agreement on both sides makes applications extremely vulnerable to changes; the client must have a very deep a-priori knowledge of:

- the web service
- the actions that can be triggered
- the meaning of XML document tags and attributes.

If the web service changes something, the client just breaks. Few web features are used to enforce interactions between endpoints; this is just an HTTP-based RPC model. The format used by XML documents is typically SOAP [30]. However, using SOAP and XML-RPC instead of plain XML does not make any difference; it is just a detail of serialization and deserialization at endpoints.

3.3.2.2 Level 1: Resources

In order to increase the robustness of client applications against changes in the server implementation, a more convenient approach is to model interactions by targeting resources instead of services. Rather than making all our requests to a single service endpoint, individual resources are addressed (one service exposes many logical resources).

Applications that use the concept of resources rather than services are classified as Level 1. This is an improvement over Level 0 because more web-oriented concepts are introduced and exploited, but HTTP is still used as a tunneling mechanism. When working with Level 1 applications, action names and parameters are typically mapped directly to a URI, rather than embedded in the semantics of XML/SOAP payloads (Figure 3.7). The action is triggered by sending an HTTP GET or POST request to the targeted URI, for example GET `http://example.com/people/123?action=delete`. One important benefit of using URIs is that resources are globally addressable. However, this is still a form RPC: the difference is that semantics are inserted in the URI, which still introduces a coupling between client and server. Moreover, the use of the GET method for operations that might create side-effects on the server (such as creating or deleting resources) does not comply with the principles defined in the HTTP specification [2], which states that GET requests should be idempotent (meaning that multiple requests have the same effect as a single request), and is therefore bad practice.

3.3.2.3 Level 2: HTTP Verbs

As we have just seen, Level 1 applications, which embed the semantics of the action to execute directly in the URI and use a single HTTP method for requests, violate the HTTP specification. A much better

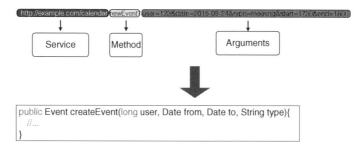

Figure 3.7 URI-to-action mapping.

Table 3.1 Semantics and effects of HTTP methods for manipulating resources.

HTTP method	Safe	Idempotent	Action
GET	yes	yes	Retrieve a resource with the given URI
POST	no	no	Create a resource (a new URI is returned by the server)
PUT	no	yes	Update the targeted resource
DELETE	no	yes	Remove the targeted resource

approach is to move the semantics from the URI to HTTP verbs when manipulating resources. In this case we refer to Level 2 applications. Resources are still addressable using URIs, and each resource can be manipulated using HTTP methods. Each method has a particular meaning and maps to a specific CRUD (create-read-update-delete) operation, as shown in Table 3.1. According to RFC 2616, HTTP methods can be safe and/or idempotent. Safe means that they have no effect on the resource (the resource remains the same). Idempotent means that the same request can be executed multiple times with the same effect as executing it once.

HTTP method semantics Level 2 applications define the following procedures for CRUD operations.

- *Creating a resource*: When creating a resource (Figure 3.8), the client must issue a POST request to the target URI of the creator of the resource or the URI of the resource to be created itself (in case this is allowed by the server). The request body might contain the initial representation (state) of the resource. The response body contains

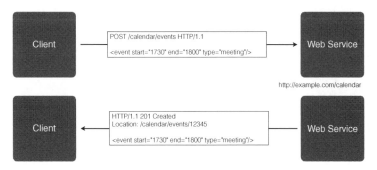

Figure 3.8 Resource creation.

the state of the newly created resource. If the URI of the newly created resource is defined by the server, it is returned in the Location header.

- *Retrieving a resource*: When retrieving a resource (Figure 3.9), the client must issue a GET request to the target URI of the selected resource. The request body is left empty. The response body contains the current state of the resource.

- *Updating a resource*: When changing a resource (Figure 3.10), the client must issue a PUT request to the target URI of the resource to manipulate. The URI is the one returned in the Location header after an initial POST. The request body might contain the new state. The response body contains the updated state.

- *Deleting a resource*: When deleting a resource (Figure 3.11), the client must issue a DELETE request to the target URI of the resource to delete. The request body is left empty. The response body might be empty or contain the state of the deleted resource.

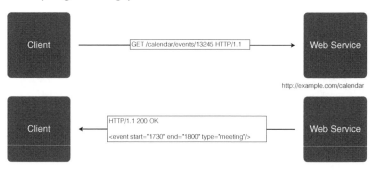

Figure 3.9 Resource retrieval.

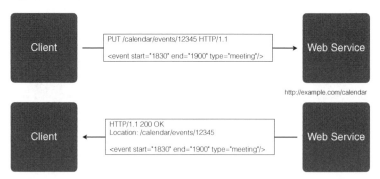

Figure 3.10 Resource update.

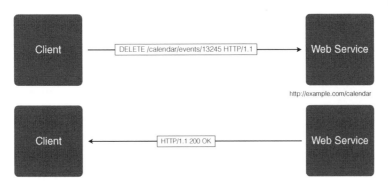

Figure 3.11 Resource deletion.

HTTP Response Code Semantics

Status (response) codes also have semantics: they have particular meaning. Each method has a specific set of status codes. For instance, 2xx indicates success, 4xx indicates a client error (e.g., 405 Method not allowed), and 5xx indicates a server error (e.g., 500 Internal server error). A client error instructs the client not to repeat the request. It can be due to a bad syntax, missing credentials or authorization, or a failure to find the resource for the given URI. A server error means that the client might choose to repeat the request since the failure was on the server side. The semantics of the most common success and failure response codes are reported in Tables 3.2 and 3.3, respectively.

Table 3.2 Semantics of 2xx HTTP status codes when manipulating resources.

Method	Status code	Reason
GET	200	OK
POST	201	Created
PUT	200	OK
	204	No content
DELETE	200	OK
	202	Accepted
	204	No content

Table 3.3 Semantics HTTP error status codes when manipulating resources.

Status code	Reason	Code
400	Bad request	The client isseud a malformed request
401	Unauthorized	The client must authenticate before performing the request
403	Forbidden	The client does not have the privileges to access the resource
404	Not found	No resource with the given URI was found
405	Method not allowed	The resource cannot be manipulated using the HTTP method
500	Internal server error	The server failed to process the request

Describing Level 2 Applications

With Level 2 applications, HTTP is no longer used just as a transport for requests, but instead is also used to describe what manipulation on the resource is being requested. By doing so, requests and response are fully descriptive and do not rely on any specific a-priori knowledge between endpoints regarding the contents of the message or how to construct a URI. Requests can therefore be read as "(READ, CREATE, UPDATE, DELETE) the resource identified by the URI."

Level 2 applications use:

- HTTP verbs and status codes to coordinate interactions and manipulate resources;
- HTTP headers to convey information (e.g., the Location header to indicate the URI of a created resource).

However, there is still some coupling between client and server applications: the client must know the URI of a resource and which methods can be invoked. Some sort of documentation is needed to let clients learn about the allowable manipulations for a resource; that is, what can be done with a resource.

The solution to this problem is to use a Web Application Description Language[1] (WADL) document [31]. A WADL document is a static description used to advertise the endpoints, the methods, and the

---

1 http://www.w3.org/Submission/wadl/.

representation formats of the resources hosted by a web application. WADL documents describe:

- sets of resources;
- relationships between resources;
- methods that can be applied to each resource, together with expected input/output and formats;
- resource representation formats (MIME types and data schemas).

WADL documents are needed to answer the following questions:

- "What actions can I take with a URI that is being linked in my representation?"
- "What is the result of such actions and how should the returned state be interpreted?"

3.3.2.4 Level 3: Hypermedia

In order to be fully compliant with the REST principles, Level 3 applications must support the *HATEOAS* principle (see Section 3.3.1.7). As we have seen, this stands for "hypermedia as the engine of application state," and is the ultimate guarantee that client and server applications are fully decoupled.

According to HATEOAS, representations of resource should be hypermedia; that is, representations contain URI links to other resources. Hypermedia embed links to drive application states. Web linking [12] specifies relation types for web links, and defines a registry for these relations. Because of this, client applications do not need any a-priori knowledge of the server application since the transitions to next possible states are embedded in the state itself. When clients reach a state of the application, the representation of the resource has a double goal:

- it describes the current state;
- it includes link information to drive the client perform the next intended transitions, according to what the server expects.

The state of a resource is the aggregation of:

- *data*: values of information items belonging to that resource;
- *links*: representing transitions to possible future states of the current resource.

This idea of letting the server drive the client through all the states of the application allows client applications to be extremely robust against changes of the server since all a client needs to do is to follow the links included in the representation of the current resource. All the client needs to know in advance is an initial URI (entry point of the application) in order to let the application reach the first state. Put in another way, the more a client ignores details about the server, the more robust and open to system evolution it is.

A direct consequence of the HATEOAS principle is the fact that hypermedia plays a fundamental role in the discovery and description of next states for the client. It is important to use or define (hyper)media types that are meaningful for the application. A typical rule for HATEOAS is "a REST API should spend almost all of its descriptive effort in defining the media type(s) used for representing resources and driving application state, or in defining extended relation names and/or hypertext-enabled mark-up for existing standard media types." Since hypermedia controls describe the resources and drive the application, they are the core of the application. A client capable of understanding the meaning of the hypermedia is fully autonomous in the execution of all the operations, regardless of any change on the server. A server can change the URI scheme independently without breaking clients and can introduce new functionalities (states) just by adding more links in the hypermedia. HATEOAS fully enables true independent evolution of systems.

Hypermedia formats

Even in Level 3 applications, clients still need to discover and interact with resources. When the client is presented links to other states, it must be able to interpret these links in order to decide which one to follow. A human-controlled application can rely on user interfaces and events input by users to advance, but this does not apply to machine-to-machine (M2M) communications. As argued before, hypermedia formats provide the means for interacting with a service and are therefore an integral part of the application's service contract. An extremely important detail of an application is the choice of hypermedia format at design time. This choice must ensure that the chosen format is able to convey all necessary information for client applications to progress. This is where semantics start to become important. Hypermedia formats can either be standard or domain-specific. Several of the hypermedia formats already in use on the Web – Atom (application/atom), RSS (application/rss), and XHTML (application/xhtml) – are widely used. XHTML supports hypermedia using hypermedia controls such as the <a> tag. Using standard formats has some benefits, such as widespread knowledge among system/software architects, developers, and IT engineers of the semantics of tags or fields and the availablity of software libraries and applications supporting such media types. However, not all standard hypermedia can meet the requirements of an application. Custom formats can be designed to map specific problem domains. Even though a discussion on semantics is beyond the scope of this book, hypermedia format is a very important design criterion, which should be carefully considered.

An IoT standard for hypermedia to be used in constrained environments is the CoRE Link Format [13]. The CoRE Link Format is an IoT-oriented web linking specification used to define attributes and relations that are meaningful for IoT applications. The CoRE Link Format will be considered in detail in Section 3.9.1.

3.3.2.5 The Meaning of the Levels

Level 1 handles complexity: a single large service endpoint is divided into multiple resources. Level 2 provides a standard set of verbs so as to handle operations uniformly (CRUD operations). Level 3 introduces discoverability, in order to make a protocol self-documenting.

3.4 The Web of Things

Modeling the IoT using web-oriented, RESTful principles can be a way to start to develop a global infrastructure of interconnected objects and to foster the development of scalable and robust IoT applications. The basic idea is to consider smart objects as tiny servers that implement Level 3 IoT applications using hypermedia and the CoRE Link Format. Building the IoT around the REST paradigm and modeling it according to web concepts allows reuse of all the experience gained in the decades of building the web. The Web of Things (WoT) provides an application layer that simplifies the creation of the IoT. By bringing the patterns of the web to the IoT, it will be possible to create robust applications in the long term and to build an infrastruture designed to scale indefinitely over time. WoT applications will bring to the IoT the same usability as the World Wide Web did with the Internet. The WoT will use a mix of HTTP and CoAP protocols, according to the specific application requirements and deployment scenarios. However, given the RESTful nature of these web-oriented protocols, the same patterns will be adopted, resulting in full interoperability between the web and the WoT. As discussed in Section 2.2.5.1, CoAP has been designed to map to HTTP; dedicated network elements (HTTP–CoAP proxies) able to perform protocol translation can be introduced in order to update client and server applications that natively communicate with different protocols.

3.5   Messaging Queues and Publish/Subscribe Communications

Of course, REST is not the only communication paradigm that can be used. REST has a number of advantages, as thoroughly discussed in Section 3.3, but it also requires objects to be tiny servers that must be accessible by clients. In many cases, due to limited processing capabilities and connectivity issues (such as firewall policies), this is not feasible. An alternative to the traditional synchronous request/response communication model (also known as the RPC pattern) is provided by messaging systems.

Messaging systems, or message-oriented middleware, implement an asynchronous communication model and loosely couple the senders with the consumers of messages, thus allowing for more flexibility and scalability. Compared to REST, senders do not send messages directly to specific receivers, about whom they do not have any knowledge at all. Messaging systems typically provide higher throughput than RPC systems; the former are bounded by network bandwidth while the latter are bounded by network latency. Moreover, the asynchronous nature of messaging systems prevents blocking of I/O operations, which may downgrade performance.

Messaging systems implement one of two asynchronous messaging approaches: message queues or pub/sub.

Message queues

In the message queue pattern, the sender sends a message to a queue on a server, where it is stored/persistent: the message is not erased immediately but is kept in memory until a consumer receives it. Only once delivered is the message deleted from the queue. This pattern implements asynchronous point-to-point communication, with total independence between the sender and the consumer.

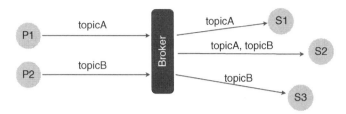

Figure 3.12 Publish/subscribe communication model.

Publish/Subscribe

In the pub/sub pattern, two kinds of entities exist: publishers and subscribers. Publishers send messages to a "topic" on the server. Subscribers can subscribe to a topic to receive a copy of all messages that have been published on that topic. This means that a message can be consumed by multiple consumers. This is similar to an application-layer multicast. Pub/sub implements the observer pattern, in an event-based paradigm.

The separation between publishers and subscribers is possible thanks to intermediary nodes, called brokers. Brokers can be implemented as message queues. Typically, the broker is involved for the following functions:

- *publishing*: publishers send messages to the broker;
- *subscriptions*: subscribers register to receive messages, possibly filtered according to some policy (content or topic).

Upon receiving a message from a publisher, the broker is responsible for dispatching messages to the subscribers, according to their subscriptions, as shown in Figure 3.12.

3.5.1 Advantages of the Pub/Sub Model

The pub/sub communication model has the following advantages:

- *Loose coupling*: Publishers need not know which subscribers receive messages or even if they exist. Contrast this with the client/server paradigm, where a client must know the URI of the server and the server must exist.
- *Scalablility*: Since brokers only need to route messages, they can be replicated easily to support higher volumes of data being transferred.

- *Lightweight implementation*: Publishers and subscribers have minimal footprints for sending and receiving messages, as most of the load is carried by the broker.

3.5.2 Disadvantages of the Pub/Sub Model

The loose coupling between publishers and subscribers has some drawbacks. As there is no direct interaction between the endpoints, the messaging contract between publishers and subscribers is inflexible. No content-type negotiation can be performed. Of course, this introduces a weakness in the architecture, as any change on the message format being published has direct consequences on the receivers. Long-term evolution of systems is therefore hard to achieve. Complex and open systems, characterized by extremely heterogeneous event semantics, can be very difficult to manage. Other drawbacks of the pub/sub model are:

- There is no support for end-to-end security between publishers and subscribers, due to the presence of the broker; messages can be encrypted but this requires encryption/decryption keys to be distributed among publishers and subscribers.
- There are throughput issues: the broker infrastructure must scale in order to avoid issues related to load peaks on both the incoming and outgoing interfaces, which can occur when the number of publishers and their publish rate and the number of subscribers increase. This can introduce slowdowns in message delivery.

3.5.3 Message Queue Telemetry Transport

Message Queue Telemetry Transport (MQTT[2]) is a lightweight, open-source, TCP-based pub/sub protocol. MQTT is a standard of the Organization for the Advancement of Structured Information Standards. The current version of MQTT is 3.1.1.

MQTT targets environments in which devices and networks are constrained. It can be used in scenarios where two-way communications between endpoints operating in unreliable networks must occur. The lightweight nature of MQTT makes it particularly suited to constrained environments where message protocol overhead and message size should be minimal.

---

2 https://mqtt.org

According to the pub/sub model, in MQTT, messages are published to a shared topic space inside the broker. Topics are used as filters on the message stream from all publishers to the broker. MQTT supports hierarchical topics in the form of a topic/sub-topic/sub-sub-topic path. Messages are delivered to all clients that have subscribed with a matching topic filter. This means that a single client can receive messages coming from multiple publishers. The matching condition is applied to the topic's hierarchy, so it possible to subscribe to just a portion of the topic. Wildcards can be used on segments of the path in order to provide finer granularity over the messages to receive.

In MQTT, the broker applies the subscription filters to the message stream it receives in order to efficiently determine to which clients a message should be dispatched. Therefore, a subscription can be considered as a conditional realtime receive operation and has a non-durable nature.

A sensor-network oriented version of MQTT, called MQTT-SN, has been defined for use in low-power and lossy networks, such as IEEE 802.15.4 networks. MQTT-SN has been designed to allow implementation on low-cost, battery-operated devices.

3.5.3.1 MQTT versus AMQP

MQTT and AMQP have a lot in common and also have many differences. We will briefly summarize their similarities and differences at a high-level.

Similarities between MQTT and AMQP

Both MQTT and AMQP share the following features:

- asynchronous, message queuing protocols
- based on TCP
- implementing an application-layer multicast
- use TLS for security at the transport layer
- widely available implementation for major platforms and programming languages.

Differences between MQTT and AMQP

MQTT and AMQP differ in the following ways:

- MQTT is more wire-efficient and requires less effort to implement than AMQP; it is well suited to embedded devices. AMQP is a more verbose protocol but provides greater flexibility.

- MQTT provides hierarchical topics with no persistence (stream-oriented approach); AMQP does not provide a hierarchical topic structure but offers persistent queues as a message storage facility (buffer-oriented approach).
- In MQTT, messages are published to a single global namespace. In AMQP, messages can be sent to several queues.
- In MQTT, the broker has AMQP supports transactions, while MQTT does not.

3.6 Session Initiation for the IoT

While the Constrained Application Protocol (CoAP; see Section 2.2.5.1) is intended to bring the REST paradigm to smart objects, there are many application scenarios that might benefit from the use of sessions. An example would be the exchange of data between an association of participants. Here, we introduce a lightweight session initiation protocol targeted at constrained environments. It re-uses the syntax and semantics of CoAP in order to create, modify, and terminate sessions among smart objects with minimal overhead.

3.6.1 Motivations

Beside the REST and pub/sub communication models, there are also many other applications in both constrained and non-constrained environments that might benefit from a non-request/response communication model. Some of these applications require the creation and management of a "session". The term "session" refers to any exchange of data between an association of participants. The communication will occur according to the network, transport, and application parameters negotiated during the session setup. Where there are two participants, the session may involve the sending of one or (probably) more data packets from one participant to the other, in one or both directions. Unidirectional sessions may be initiated by either the sender or the receiver. Examples of sessions in IoT scenarios may be the data flow generated by a sensor and sent to a particular recipient for further processing, or data streams exchanged by two interacting smart toys.

The Session Initiation Protocol (SIP) [5] is the standard application protocol for establishing application-layer sessions. It allows the endpoints to create, modify, and terminate any kind of (multi)media session: VoIP calls, multimedia conferences, or data communications for example. Session media are transmitted end-to-end, in a peer-to-peer fashion, either using specific application/transport layer protocols, such as RTP [16], or as raw data on top of the TCP or UDP transport protocols. SIP is a text-based protocol, similar to HTTP, and can run on top of several transport protocols, such as UDP (default), TCP, or SCTP, or on top of secure transport protocol such as TLS and DTLS. Session parameters are exchanged as SIP message payloads; a standard protocol used for this purpose is the Session Description Protocol [17]. The SIP protocol also supports intermediate network elements, which are used to allow endpoint registration and session establishment. Examples include SIP proxy servers and registrar servers. SIP also defines the concepts of transaction, dialog, and call as groups of related messages, at different abstraction layers.

Although SIP was defined for Internet applications, it has the potential to be reused in the IoT. However, this would require suitable adaptation in order to meet the requirements of constrained environments: the large message size and the processing load for the parsing of standard text-based SIP messages is not suitable for constrained environments.

For these reasons, in this section, we propose an alternative mechanism for session initiation, based on CoAP, aiming at allowing constrained devices to instantiate communication sessions in a lightweight and standardized fashion. Session instantiation can include a negotiation phase, in which the parameters used for all subsequent communication are determined. We propose to reuse not only the CoAP message format, but also the REST paradigm (and methods) provided by CoAP for initiating sessions, rather than compressing SIP into a binary protocol and redefining new methods. This approach brings several benefits, such as the possibility of taking advantage of the well-known REST paradigm and existing implementations of CoAP and avoiding the need to include other software libraries in smart objects, which are typically very limited in available memory.

3.6.2 Lightweight Sessions in the IoT

The reuse of CoAP's message syntax, rather than defining a brand new ad-hoc session initiation protocol for constrained environments, has several motivations:

- Since CoAP maps HTTP methods well, and since SIP is very similar to HTTP, the message format of CoAP is well-suited to carrying information required by a protocol like SIP.
- Since CoAP already has many implementations in constrained operating systems, such as Erbium [32] for Contiki OS, its use would not require new implementation and testing efforts.
- The possibility of using the same software libraries as CoAP allows the memory footprint to be kept very small – a critical issue for smart objects with very limited available memory.

There are two possible approaches for establishing, managing, and tearing down a session when using the message format defined by CoAP:

- defining new ad-hoc methods for session initiation together with their own semantics, encoded according to CoAP syntax;
- reusing the semantics of CoAP methods and trying to manage sessions in a RESTful fashion.

Our opinion is that the latter approach is preferable since it would maximize the reuse of CoAP: both the message format and the semantics. This approach would also being the benefit of avoiding changes to the standard or introducing a new one, as would be needed if new methods were to be defined.

Treating the problem of session initiation in a RESTful fashion poses some issues, mainly related to the concept of a session:

- Is a session compatible with the RESTful paradigm?
- In RESTful terms, how should a session be defined?
- What are the roles involved in the management of a session?
- Can a resource be negotiated among two or more parties?

The answers to these questions drive the design of a suitable lightweight session initiation protocol. A first consideration is that a unidirectional session can be considered as a flow of data from a data source and to a destination. A bidirectional session (i.e., a session where each of the endpoints of communication generate independent – but related – flows of data) can therefore be considered as the union of two unidirectional sessions. This idea can be extended to a session among multiple parties, which can be seen as the union of several unidirectional sessions.

A second consideration is related to the role of the endpoints involved in the sessions. In SIP, an endpoint, termed a user agent (UA), can act in one of two roles when participating in a session:

- a user agent client (UAC), which sends SIP requests (typically to initiate a session, through INVITE requests);
- a user agent server (UAS), which receives requests and returns responses.

SIP UAs typically act as both UAC and UAS because, in general, it cannot be determined in advance whether a UA will initiate a multimedia session (caller) or respond to a request for initiating a session (callee). However, in constrained IoT scenarios, it can be assumed that, because of their nature, smart objects will either as a UAC or, more often, as a UAS. Note that the role in establishing the session (UAC or UAS) is not strictly related to the role during the session (sending or receiving data). It is important to note that a session does not necessarily imply a multimedia flow, but it is related to a long-lasting flow of data that adheres to some parameters that can be negotiated among the endpoints during the initiation phase. This feature is obviously compatible with duty-cycled devices.

In RESTful terms, a session can be considered a "resource" containing all the parameters related to the flow of data. The session identifier is the URI of the resource. Figure 3.13 shows an example of an XML-based representation of a session resource, based on SDP. Other hypermedia formats can be also used as the resource representation, for instance based on JSON. The proposed protocol makes no assumption about the actual hypermedia format used to represent resources.

Figure 3.13 Example of an XML-based representation of a session resource carrying SDP session information.

```
<?xmL version="1.0" encoding="UTF-8"?>
<cosip:session
    xmlns: cosip="http://schema.org/ietf/cosip"
    xmlns: sdp="http://schema.org/ietf/sdp">
    <cosip:params>
        <sdp:c>In IP6 aaaa::2</sdp:c>
        <sdp:m>audio 4000 RTP/AVP 0</sdp:m>
        <sdp:a>rtpmap:0 PCMU/8000</sdp:a>
    </cosip:params>
</cosip:session>
```

3.6.2.1   A Protocol for Constrained Session Initiation

In this section, we present the proposed procedures for establishing, maintaining, and terminating sessions, according to the REST paradigm and to the discussion about sessions and resources in Section 3.6.2.

3.6.2.2   Session Initiation

According to the well-known session setup model used by several session initiation and call-setup signaling protocols (e.g., SIP, H.323, ISUP), the UAC (the caller) starts by sending the first request message to establish a session. Then, the UAS (the callee) can respond, either accepting or refusing the request. A positive response is then confirmed by the UAC in order to inform the UAS that the response has been received and the session can start. During this three-way handshake the UAC and UAS may negotiate the session by specifying session-related attributes, such as data type, flow direction (from UAC to UAS, from UAS to UAC, or bidirectional), flow endpoints, application and transport protocols, formats, and encoding parameters. Session negotiation is typically implemented through an offer/answer model, where one party sends an "offer" and the other party sends an "answer". Although some protocols (e.g., SIP) support both UAC- or UAS-initiated negotiation (i.e., both UAC and UAS may send the offer, and the other party sends the answer in the next message), session negotiation is usually initiated by the UAC.

According to the REST paradigm, a session is a resource maintained by the server. In this case, the client (UAC) requests from the server (UAS) the creation of such a resource; the server may either accept the request and create the resource (thus acting as an origin server) or refuse it. The session setup procedure is shown in Figure 3.14. The UAC sends a CoAP CON POST request targeted to the UAS, including a payload containing the offer. This is encoded in a suitable format, such as SDP, XML, or JSON. The POST request also contains a CoAP *Observe* option set to 0 (i.e., register) in order to specify at the same time that the UAC is also interested in observing the created resource. This behavior is needed in order to support server-side session modifications and termination, as described in Sections 3.6.2.3 and 3.6.2.4. Then, the UAS may:

- accept the session, by replying with a CoAP CON 2.01 "Created" response, including the new session identifier (resource URI)

Figure 3.14 Session initiation procedure.

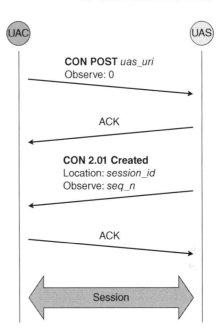

within a *Location-Path* CoAP option field and the answer within the payload, reporting the result of the session negotiation (i.e., the representation of the created resource);

- refuse the session.

In both the request and the response, the format of the session description (the offer and the answer) is specified by the *Content-Format* CoAP option field.

In order to accomplish reliable transmission in the session setup phase, the CoAP CON POST request and the CoAP CON 2.01 responses should be acknowledged. CoAP defines also mechanisms to encapsulate the response in an ACK message (piggy-backing), in order to reduce the number of transmitted messages. In Figure 3.14, the UAS confirms the CON POST with an ACK, then sends the CON 2.01 response, confirmed by the UAC with an ACK message. In this case, the setup procedure involves four messages. although retransmissions might occur in the event of message losses or timeouts. When the setup procedure has completed, both parties may start sending the data according to the negotiated session.

3.6.2.3   Session Tear-down

At the end of a session, the UAC or the UAS might wish to explicitly communicate to the other party the intention of ending the data communication. Even though this is not strictly required, it is preferable to perform this "graceful termination" procedure in order to inform the other party to stop listening for incoming data and to free the allocated processing and/or memory resources. Although this could be done directly within the data plane, using specific features of the data communication protocol (when available), it is usually implemented at the control plane using the same signaling protocol adopted for session setup.

For the proposed CoAP-based session initiation protocol, the tear-down procedure can be initiated either by the UAC or the UAS. If the UAC wants to tear down the current session, it starts the session-termination procedure by sending a CoAP CON DELETE request targeted to the session resource URI (previously obtained by the server during the setup within the 2.01 response). The UAS will process the received DELETE request: if a corresponding active session is found, the session is terminated and a CoAP CON 2.02 "Deleted" response is sent. According to the CoAP *Observe* option extension, if the UAS is willing to tear down the session, it can simply send a CoAP CON 4.04 "Not found" response matching the same *Token* option included in the original POST request that created the resource (session). Then the UAC confirms the reception of the 4.04 response through an ACK message.

Figure 3.15 shows the session tear-down procedure, with both UAC- and UAS-terminated sessions depicted.

3.6.2.4   Session Modification

We consider also the case in which the UAC or the UAS may want to change the current active session. The session change may involve the modification of session directions, data types, protocols, or other session parameters.

If the UAC wants to change the session, it sends a CoAP CON PUT request targeted to the session resource URI. The UAC will include the offer for the new changed session in the message payload. If the UAS agrees on the new session description, it sends back a CoAP CON 2.04 "Changed" response containing the new representation (i.e., the answer describing the changed session). The CON 2.04 is

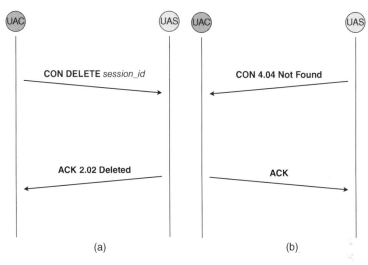

Figure 3.15 Session terminated by (a) UAC and (b) UAS.

then confirmed by the UAC and the session modification can be applied by both endpoints.

If the UAS wants to change the session, a new answer (compatible with the offer provided by the UAC during the setup) should be sent. According to the CoAP *Observe* option extension, this is performed by sending a new CoAP CON 2.04 "Changed" response reporting the same *Token* option included in the original POST request that created the session. Figure 3.16 shows the UAC- and UAS-initiated session modification procedures.

3.7 Performance Evaluation

3.7.1 Implementation

In order to provide a proof-of-concept of the proposed session initiation protocol and to evaluate and compare its performance with other mechanisms, the proposed protocol and some testing applications have been implemented. The implementation is based on Java, due to its simplicity, cross-platform support, and the availability of a CoAP library [23], which has been suitably extended with the

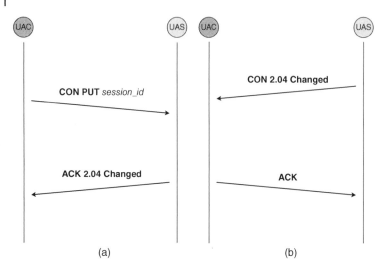

Figure 3.16 Session updated by (a) UAC and (b) UAS.

session initiation mechanisms described in Section 3.6.2.1. As a proof-of-concept, we implemented two reference applications:

- a CoAP UA, which allows the establishment of end-to-end audio sessions;
- a CoAP server, which provides a simple data retrieval service, with periodic data notification.

The first application consists of an UA, including both UAC and UAS components, which allows establishment and tear down of unidirectional or bidirectional RTP audio flows between the two UAs. This application is based on a standard SIP UA [33], modified in order to handle session setup according to the proposed protocol. The audio session (i.e., the RTP flows) is negotiated through the standard SDP protocol using the classical offer/answer model.

The second application is a simple server that allows a client to initiate a data retrieval service (for example, to receive values provided by a sensor) by establishing a data session with a server and specifying the encoding data format and the receiver socket address. Raw data are then encapsulated and sent in UDP packets.

The source code of the extended CoAP library, together with the two reference applications, is freely available [34].

3.7.2 Experimental Results

We conducted experimental evaluations in order to compare the proposed solution with other protocols and mechanisms. We first compared our protocol with other standard and non-standard session initiation signaling protocols by running the same UA (suitably modified) using three different underlying session setup protocols:

- standard SIP;
- a compressed version of SIP that encodes SIP signaling in a CoAP-compatible binary message format (reusing CoAP syntax) [35];
- the new CoAP-based session initiation protocol proposed here.

The evaluation results are reported in Figure 3.17. The average size per message of the exchange are reported for the three types of signaling protocol. In all three cases, SDP was used for the session negotiation. In particular, an SDP offer (115 B) was included within the SIP INVITE and the CoAP POST requests. The SDP answer (115 B) was included in the SIP 200 "OK" and CoAP 2.01 "Created" responses.

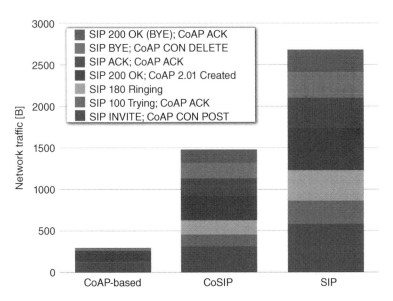

Figure 3.17 Network traffic comparison between three session initiation mechanisms: RESTful (CoAP-based), compressed SIP (re-using CoAP-syntax), standard SIP.

Table 3.4 Comparison of different session initiation protocols.

Protocol	Signaling only		Signaling plus SDP	
	Network traffic [B]	Ratio over SIP	Network traffic [B]	Ratio over SIP
SIP	2433	1	2663	1
Compressed SIP	1249	0.51	1479	0.55
CoAP	65	0.027	295	0.11

Table 3.4 shows the total number of bytes transmitted in the exchange for the three signaling protocols, comparing the network traffic without and with SDP payload, and the ratio of the traffic volume compared to standard SIP for each.

It is possible to note that, while standard SIP requires the transmission of 2663 B (2433 B for SIP signaling only), the proposed CoAP-based approaches requires only 295 B to be transmitted (only 65 B for CoAP signaling), representing a large decrease. In fact, the ratio between the proposed approach and SIP is only 0.027, so the gain is significant. The gain, in terms of overall network traffic, is due to:

- the low-overhead of binary CoAP compared to the verbosity of text-based SIP;
- the design of SIP, which is mainly intended for rich multimedia applications and not for IoT scenarios.

The proposed solution is better suited to meeting the strict requirements of constrained environments, which force the adoption of mechanisms with low overhead in order to minimize energy consumption and to avoid delays due to retransmission (which may occur when operating in LLNs).

A second comparison was conducted in order to evaluate the two different CoAP-based approaches for retrieving data from a remote server. We considered a client interested in receiving data from a remote server (say, a sensor periodically providing value updates). The two approaches considered are:

- The CoAP client registers itself with the server as an observer of the resource (i.e., using the *Observe* option) and periodically receives updates of the new resource state according to the CoAP resource observing model [36] through CoAP NON 2.04 "Changed" responses.

- The CoAP client establishes a session with server, negotiating both the receiver socket address and the data encoding format; all data values are then sent by the server within UDP packets.

Figure 3.18 shows the results in terms of overall generated network traffic as a function of the number of exchanged packets. The establishment of the session requires more bytes at the beginning of the communication. However, after a few data packets, the solution based on a CoAP-negotiated session outperforms the solution based on CoAP observing in terms of overall network traffic. In the tested application, each update consisted in 4 B of data. CoAP observing introduces an additional 7 B overhead, resulting in a trade-off point at just after 7 packets. As for the tear-down procedure, the session-based solution is comparable to session termination through resource observing, so there is no additional overhead.

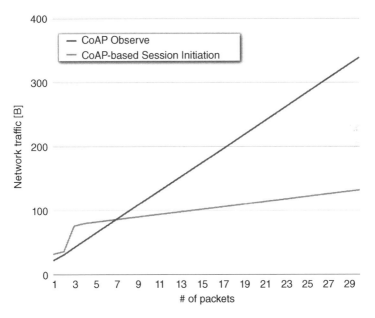

Figure 3.18 Network traffic comparison between i) CoAP observing and ii) session-based data flow as a function of the number of exchanged packets (adimensional).

Table 3.5 Comparison of Java bytecode footprints.

Library	Support	Size [B]
mjCoAP	RFC 7252 draft-ietf-core-block draft-ietf-core-observe	165227
mjCoAP + sessions	RFC 7252 draft-ietf-core-block draft-ietf-core-observe session initiation	177661

As a final performance metric, the memory footprint introduced by session establishment logic is considered. Table 3.5 shows a the comparison of bytecode footprint for the Java libraries (the jar files) with and without the session support. The increase of the size is negligible.

3.7.3 Conclusions

Many constrained and non-constrained applications might benefit from communication models other than REST and pub/sub, and might require the creation and management of a session (i.e., any exchange of data between an association of participants). Some session initiation protocols, such as SIP, have been introduced but SIP cannot simply be reused in the IoT because of its large overhead which mean that it cannot meet the requirements of constrained environments. For this reason, a lightweight protocol for session initiation, reusing the syntax and semantics of CoAP, and considering a session as a CoAP resource, can be introduced. In fact, the proposed approach for session establishment, management, and termination is, by nature, standard and adheres to the REST paradigm. It provides several advantages such as low-overhead and reuse of existing implementations (thus, minimizing memory footprint of smart objects).

A Java implementation of the proposed protocol has been used to conduct a performance evaluation, aiming at highlighting the advantages, in terms of network traffic (and consequently, on energy-consumption), over alternative solutions, such as SIP and CoAP observing. The results show that the proposed lightweight session initiation protocol meets the IoT's low-overhead requirements and that applications that rely on this approach outperform others, based on different mechanisms, such as SIP sessions and CoAP observing.

3.8 Optimized Communications: the Dual-network Management Protocol

The trade-off between high-performance data transmission and low-energy consumption is a long-standing issue in the field of wireless communications. Efficient data exchange is critical for the battery-powered mobile devices typically used in distributed surveillance scenarios, which may be required to transmit video or audio streams between in-network nodes. In such cases, the devices carried by operators must be as lightweight as possible. However, LLNs do not provide sufficient bandwidth to meet the requirements of streaming transmissions, while high-performance communications result in high energy consumption, shortening the lifetimes of the devices. Here we present the data-driven IoT-oriented Dual-network Management Protocol (DNMP), which uses two IP-addressable radio interfaces on the same node: one with low energy consumption (and throughput) and one with high throughput (and energy consumption). The low-power network acts both as an independent data plane and as a control plane for the high-throughput network, the latter being turned on whenever necessary to support multimedia streaming. In order to validate the protocol, we consider the integration of a low-power IEEE 802.15.4 radio transceiver and a high-throughput IEEE 802.11s radio transceiver. An extensive experimental investigation is then carried out, with the following goals:

- investigating the performance of the two overlaid networks (IEEE 802.15.4 and IEEE 802.11s);
- determining the critical threshold, in terms of amount of data to be transmitted, beyond which the use of DNMP becomes advantageous.

3.8.1 DNMP Motivations

There are many application fields for IoT devices, ranging from remote environmental monitoring to smart surveillance. Clearly, the performance requirements vary significantly as well. For instance, in a remote monitoring scenario [37], sensor nodes are statically deployed and may send only a few data packets per day to a sink node; a few packet losses are likely to be tolerable. In a smart surveillance scenario [38], on the other hand, nodes must be able to send video or audio streams on a mesh/ad-hoc network: resilience and very low packet-loss rates are key system requirements. In particular, many surveillance scenarios require the transmission of audio/video streams between two or more mobile devices: for example, policemen or firemen can take advantage of this capability to increase their context awareness during operations. However, in order to maximize the lifetime of the equipment without burdening the operators with heavy batteries, it is important to provide mechanisms that minimize energy consumption.

The IEEE has developed several standards that partially meet the different requirements of different IoT scenarios. For instance, the IEEE 802.15.4 standard promises to make the implementation of systems with years of activity feasible thanks to the adoption of low-power consumption hardware. In this context, the IETF ROLL Working Group has introduced RPL (IPv6 Routing Protocol for Low-Power and Lossy Networks) [39], a lightweight routing protocol for LLNs. In addition, the research community has developed operating systems for constrained objects. Among these, we should mention ContikiOS [40], which, besides supporting low-power standards like 6LoWPAN and RPL, allows for full IP networking with standard IP protocols such as UDP, TCP, and HTTP.

The IEEE has also defined several standards to achieve high-performance communication. Among these, IEEE 802.11 is a set of specifications that provide different standards for wireless networking, especially for infrastructured wireless networking. The most commonly used versions of IEEE 802.11 (IEEE 802.11g and IEEE 802.11n [41, 42]) rely on access points to collect data from remote terminals. These standards guarantee bit rates of up to 54 Mbps in g mode or 600 Mbps in n mode. However, these solutions do not provide mesh networking, a s mode does. In particular, the IEEE has released the IEEE 802.11s amendment, which extends IEEE 802.11b/g/n [41, 42] by defining an architecture and a protocol stack supporting both broadcast/multicast and unicast delivery using "radio-aware metrics over self-configuring multi-hop topologies" [43].

In order to provide a trade-off between performance and energy efficiency, we here present the IoT-oriented data-driven energy-efficient

Dual-network Management Protocol (DNMP), which relies on the coexistence of a low-power/low-throughput (LPLT) network and a high-power/high-throughput (HPHT) network. The former transmits low-rate data (data plane) and acts as an overlay (control plane) that, according to the information coming from routing tables, selectively activates, via IP-based communications, a set of high-speed interfaces (forming a second data plane). When a large burst of data, such as a video or an audio stream, needs to be transferred, the low-power network selectively turns on the high-speed interfaces of the active nodes along the path between source and destination. When streaming has completed, the low-power network turns off the high-speed interfaces to avoid wasting energy. We should point out that DNMP is agnostic about the MAC/routing protocols in the nodes' protocol stacks and only requires that the dual radio interfaces are univocally IP addressable. In order to validate DNMP, we have evaluated its performance, in terms of overhead and latency, using the IEEE 802.15.4 standard for the low-power network (data/control plane) and the IEEE 802.11s standard for the high-throughput network (data plane) (see Section 3.8.5). The critical amount of data beyond which DNMP becomes effective is identified.

3.8.2 Related Work

The idea of using a low-power overlay to control a high-speed network has been considered in the literature. For instance, Gummeson *et al.* the authors used two low-power technologies to create a unified network [44]. In their scheme, the nodes learn about the characteristics of radio channels through exploration; then, they dynamically and continuously select the most efficient radio channel among those available. Although radios are switched at runtime, an abstraction of a unified link layer is provided to applications running on multi-radio platforms. However, there was no concept of traffic quality of service in this study. DNMP, on the other hand, clearly decouples the two networks and selects the most appropriate communication interface for the type of traffic to be transferred.

Sengul *et al.* propose using multiple radios to transmit bursts of data accumulated at sensors [45]. They highlighted the existence of a trade-off when deciding whether to rely on low-power devices or to turn on high-speed interfaces. A similar approach was proposed by Wan *et al.*, who triggered their high-speed interfaces as soon as network congestion was detected on the low-power interfaces [46]. Unlike these approaches, DNMP is more general and can be exploited for any kind of data transfer. In addition, in the Sengul and the Wan approaches [45, 46], the high-speed plane is used only as backup for the low-power plane, whilst in DNMP the two planes can be used simultaneously.

Stathopoulos *et al.* used an IEEE 802.15.4 network to switch on IEEE 802.11g interfaces [47]. DNMP can be seen as a formalized extension of this approach, since we generalize the wake-up mechanism to any pair of radio interfaces. In addition, since we utilize the presence of IP-addressable interfaces, the DNMP approach is general and agnostic of protocols running at lower layers.

The growing interest in dual-radio devices is witnessed by the efforts made by the scientific community to develop energy-efficient hardware. Jurdak *et al.* used two IEEE 802.15.4 transceivers to achieve high throughputs with reduced energy consumption [48]. In DNMP, no specific standards are considered a priori. However, as described in Section 3.8.4, as this work focuses on IoT scenarios, the validation has been carried out by integrating a TelosB mote, acting as an IEEE 802.15.4 radio interface, and an IEEE 802.11s dongle, acting as the corresponding interface, on a Raspberry Pi.

3.8.3 The DNMP Protocol

DNMP is a lightweight User Datagram Protocol (UDP)-based application-layer protocol intended to provide a simple mechanism to allow devices equipped with a LPLT radio interface and a HPHT radio interface to establish efficient end-to-end routes between pairs of nodes over a multi-hop network.

The LPLT radio interface is managed by an IP-based communication protocol, denoted as RP_{low}. DNMP exploits RP_{low} as a control-plane protocol to setup one or more routes from a given source to an intended destination, using the HPHT interface, which, in turn, is managed by a data-plane protocol, denoted as RP_{high}. Depending on the user application needs, the DNMP control plane RP_{low} switches on and off the HPHT interfaces. This represents an RP_{high} overlay that can be used when there are transmissions with high-throughput requirements (e.g., multimedia streaming), for which LPLT interfaces are unsuitable.

According to DNMP, each node keeps a counter, referred to as C_{high}, of the number of RP_{high} routes that require its HPHT interface to be active to ensure end-to-end communication between the requesting source/destination pairs. LOCK and UNLOCK messages are used, respectively, to increment and decrement each involved C_{high} by 1. The value of a node's C_{high} depends on the number of concurrent routes that can be set up; more precisely, if a node is needed for n different routes, its C_{high} value is n. When a node's C_{high} reaches 0, this node can switch its RP_{high} interface off because it is no longer needed in any route. DNMP is based on a request/response communication model and defines the syntax of request and response messages that are used to map the LOCK and UNLOCK operations. As DNMP is UDP-based, it provides mechanisms for reliable message transmission, specifying retransmission of requests that did not receive a response, with an exponential backoff.

We note that according to DNMP, the RP_{low} protocol can be used, not only as a control-plane protocol, but also as a data-plane protocol for transmissions requiring a bit-rate below a certain threshold (i.e., not requiring the throughput provided by RP_{high}). The role of RP_{low} is left to the application, which can decide whether to use DNMP (thus dynamically switching between RP_{low} and RP_{high}) or not, depending on the bit-rate requirements of the data transmission.

DNMP message format
DNMP messages are encoded in a simple binary format. Each message starts with five fixed-size fields, as specified in Table 3.6. These are followed by a variable number of "Hop ID" message fields.

Route establishment
DNMP sets up RP_{high} routes in a three-phase process, as shown in Figures 3.19–3.21. The first stage of the process, called "locking", switches on all RP_{high} radio interfaces of the devices that are on the RP_{low} route from a given source to a specific destination, using a LOCK message. After the locking, the route identification stage determines which DNMP nodes are actually needed for the multi-hop communication over RP_{high}. Finally, the unlocking stage can be used to switch off, using UNLOCK messages, all the RP_{high} radio interfaces that are not needed for the end-to-end communication among the specified endpoints. This prevents unnecessary energy consumption by the RP_{high} radio interface.

Locking phase When a source node, denoted as S, needs more bandwidth for transmitting data to a destination node, denoted as D,

Table 3.6 DNMP message fields.

Field	Length (bits)	Description
Version	2	Indicates the DNMP version number.
Code	6	Split into a 3-bit class (most significant bits) and a 3-bit detail (least significant bits), documented as c.dd, where c and dd are two digits from 0 to 7.
		The class can indicate:
		• a request $(0 = (000)_2)$,
		• a success response $(2 = (010)_2)$,
		• a client error response $(4 = (100)_2)$,
		• a server error response $(5 = (101)_2)$.
Message ID	16	Used to detect duplicate messages and to match request/response messages.
Sender ID	8	The RP_{low} short address of the originator of the request, i.e., the node willing to setup a RP_{high} route to another node.
Destination ID	8	The RP_{low} short address of the target of the request, i.e., the node the RP_{high} route needs to reach.
Hop ID	8	The RP_{low} short address of the intermediate node that has received the request. This field is used for route-tracing purposes over RP_{low}.
		By reporting the list of traversed nodes, the sender can learn about which nodes have increased their RP_{high} interface counter.

it sends a LOCK request through the RP_{low} control-plane protocol, to select the next-hop node R_1 (at the network layer) in the direction towards D by exploiting its RP_{high} routing table to extract R_1's IP address. On receiving the LOCK request, R_1 increments its C_{high} by 1 and, if previously off, switches its RP_{high} interface on. Next, R_1 appends its identifier to the tail of the request message, in order to trace that it has received and processed the message. Finally, R_1 selects the receiving node R_2 in the next hop and relays the updated request message. These steps are repeated until the LOCK request reaches D. The request message received by D contains all the identifiers of the

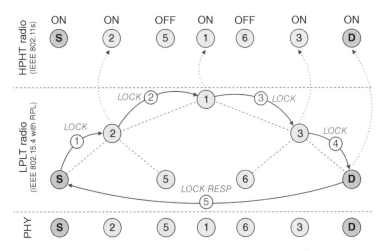

Figure 3.19 DNMP message flow for the RP_{high} route establishment: locking.

nodes that relayed the message. At this point, D sends a response message to S as an acknowledgement, including the identifiers of all the intermediate nodes so that S becomes aware of the route that the request has followed. This procedure is shown in Figure 3.19.

Route identification phase At the end of the locking phase, the LOCK request issued by S has reached D, and all intermediate nodes $\{R_i\}$ have incremented their C_{high}. S can now issue a trace request for D in order to learn which nodes are actually needed to route a message to D using RP_{high}. The nodes that are not needed for the RP_{high} multihop communication between S and D are then put in a list, called the unlock list (UL), while nodes that are needed are put in another list, called the locked list (LL). The main steps of this phase are shown in Figure 3.20.

Unlocking phase Finally, S sends an UNLOCK request over RP_{low} to all nodes in the UL, which have increased their C_{high} but are not needed for the end-to-end communication between S and D. After receiving the UNLOCK request, an intermediate node decrements its C_{high} by 1 and, if C_{high} reaches 0, switches off its RP_{high} interface to save energy. After doing so, the node sends a response back to S to acknowledge that it has received and processed the request. The steps of the unlocking phase are shown in Figure 3.21.

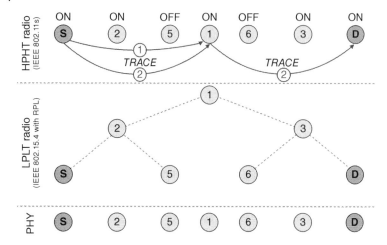

Figure 3.20 DNMP message flow for the RP_{high} route establishment: route identification.

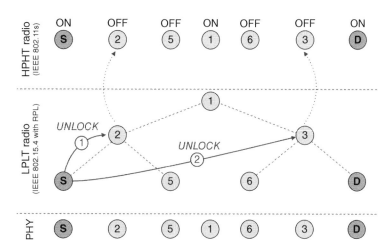

Figure 3.21 DNMP message flow for the RP_{high} route establishment: unlocking.

Route tear-down

When S completes its RP_{high} data transmission to D, it sends UNLOCK requests to the nodes that are in the LL, so that each of these nodes decreases by 1 the value of C_{high} and, if possible, turns off its HPHT interface.

3.8.4 Implementation with IEEE 802.15.4 and IEEE 802.11s

In order to validate DNMP, we considered an IEEE 802.15.4 LPLT interface and an IEEE 802.11s HPHT interface. In the following, we first provide more details on the two standards and then we characterize their DNMP-based integration on a node.

3.8.4.1 LPLT Networking

The IEEE 802.15.4 standard, which defines physical (PHY) and medium access control (MAC) layers for wireless personal area networks (WPANs), targets reduced energy consumption and ubiquitous communications. Although IEEE 802.15.4 can operate on three ISM frequency bands, here we use the band at 2.4 GHz since it offers the largest channel bandwidth and allows for bit rates of 250 kpbs. Although the maximum transmit power defined by the standard is around 0 dBm, the transmission range is about 100 m with currently available transceivers. At the MAC layer, the IEEE 802.15.4 standard uses a carrier sense multiple access with collision avoidance (CSMA/CA) strategy to minimize energy dissipation. Since the implementation of the LPLT network is carried out via a TelosB running the Contiki OS (an open-source operating system for low-power nodes) [40], we use the ContikiMAC protocol [49], which allows us to achieve useful energy savings by cyclically turning on and off radio interfaces. In order to provide full in-network connectivity, we rely on RPL, which is emerging as the *de facto* routing protocol for constrained networks. RPL allows us to build a multi-hop and dynamically reconfigurable routing tree between each pair of nodes in the network. Finally, at the transport layer, we use UDP, since it is more suitable for constrained nodes than traditional TCP.

3.8.4.2 HPHT Networking

The IEEE 802.11s standard is an amendment of the IEEE 802.11 standard and is fully compliant with the IEEE 802.11b/g/n standards. While IEEE 802.11b/g/n networks operate only in infrastructure mode – that is, all communications are centralized – the IEEE 802.11s standard overcomes this limitation to enable mesh networking. The maximum transmit power allowed by the standard is 20 dBm, which is 100 times higher than that used by IEEE 802.15.4. Due to its inherent dependence on IEEE 802.11n, IEEE 802.11s operates at a maximum net data rate ranging from 54 to 600 Mbps (when it uses 40 MHz channels and is configured to support multiple-input multiple-output communications and frame aggregation).

For medium access, mesh stations implement the mesh coordination function (MCF). MCF relies on the contention-based protocol known as enhanced distributed channel access (EDCA), which is itself an improved variant of the basic IEEE 802.11 distributed coordination function (DCF). Using DCF, a station transmits a single frame of arbitrary length. With EDCA, a station may transmit multiple frames whose total transmission duration may not exceed the so-called transmission opportunity limit. The intended receiver acknowledges any successful frame reception. The default path selection protocol foreseen by the IEEE 802.11s standard, called the hybrid wireless mesh protocol (HWMP), combines the concurrent operations of a proactive tree-oriented approach with an on-demand distributed path selection protocol derived from the ad-hoc on-demand distance vector protocol. The proactive mode requires a mesh station to be configured as a root mesh station.

3.8.4.3 Node Integration

The two reference (LPLT and HPHT) standards were integrated on a single node in order to deploy a real testbed for dual-interface nodes. Regarding the IEEE 802.15.4 standard, the chosen node is a SkyMote TelosB. TelosB is the best known mote, and is a verified hardware platform that uses Contiki OS. According to the CC2420 datasheet,[3] the mote consumes 50 mW in transmission/reception phases and $0.24 \, \mu$W in sleep mode. As for the IEEE 802.11s standard, the choice of the USB dongle was not straightforward, because compatibility with the IEEE 802.11s draft strictly depends on the available drivers for the chosen platform. Our choice fell on the TP-LINK TL-WN722n USB dongle, as there is open source firmware for its chipset (the Atheros AR9271). This firmware can be compiled on different platforms and natively supports the mesh point mode. The power consumption of the TP-LINK IEEE 802.11s interface is 1.1 W in transmission mode and 0.2 W in reception mode. The model used operates at a voltage of 3.3 V and has no power consumption in sleep mode.[4] The selected radio interfaces have then been integrated on a Linux-powered Raspberry Pi rev.B, whose two USB ports have been used to connect the

---

3 http://inst.eecs.berkeley.edu/\HCode{<SPitie/>}cs150/Documents/CC2420.pdf.
4 https://wikidevi.com/wiki/Atheros_AR9271.

Figure 3.22 Picture of the integrated node. A Raspberry Pi hosts a TelosB mote as IEEE 802.15.4 interface and a TP-LINK as IEEE 802.11s interface.

TelosB and the Wi-Fi dongle. The integrated dual-interface node is shown in Figure 3.22.

The DNMP code was developed in C for Contiki OS. This allows us to have low-level control over the IEEE 802.15.4 network, enabling us to format UDP packets correctly and to handle routing. If, according to the LOCK-UNLOCK policy, RP_{low} has to modify the state of the RP_{high} interface, Contiki OS has to speak with the Raspberry Pi and make some modifications to the actual network configuration. In order to communicate with the TelosB, we have adapted the *serialdump.c* serial console provided with Contiki OS, enabling the motes to parse the data coming from the serial port and to react appropriately to the various messages. In this way, *serialdump.c* becomes a sort of "daemon", listening on the serial port for potential messages of interest. Due to its characteristics, *serialdump.c* is also in charge of the HWMP route-identification and unlocking phases, as outlined in Section 3.8.3.

3.8.5 Performance Evaluation

In this section, we analyze the performance of the proposed overlaid dual-radio system. The chosen scenario consists of a surveillance network for infrastructure monitoring. In the presence of an alarm, high-priority data (such as a video stream) needs to be transmitted from the terminal of an operator to the terminal of another operator.

3.8.5.1 Experimental Setup

The testbed consists of a linear network of DNMP-enabled dual-radio nodes (an example of such an application would be for monitoring

First: all 802.11s interfaces on RPL route are **turned ON**

Second: Useless 802.11s interfaces are then **turned OFF**

Figure 3.23 IEEE 802.15.4 vs. IEEE 802.11s transmission ranges.

the perimeter of a building). Using the maximum transmit power, the transmission range of the IEEE 802.11s dongles is nearly twice the transmission range of IEEE 802.15.4 TelosB motes (Figure 3.23).

We refer to the scenario shown in Figures 3.19–3.21, with a sequence of five nodes. Node S wants to communicate with node D. If nodes S and D use only the IEEE 802.15.4 network, they need to create a four-hop route, since every node has radio visibility of only its closest neighbors. On the other hand, if nodes communicate through the IEEE 802.11s network, HWMP, which IEEE 802.11s uses for routing, leads to the creation of a two-hop route between the two nodes.

3.8.5.2 Operational Limitations of IEEE 802.15.4

We first investigate the features and limitations of the IEEE 802.15.4 standard. In particular, the gap, in terms of performance, between ideal Contiki nodes (simulated in the Cooja simulator) and real nodes (operating in an indoor scenario) needs to be determined. The analysis evidenced a significant difference between real and ideal nodes, in terms of packet-loss rate and average delay. In particular, in order to perform simulations of the experimental scenario, we fine-tuned Cooja. After some testing, we heuristically set the transmission and reception probabilities of each node to 0.95. By doing so, the performance of the simulated and real nodes showed the same trends. Due to the limited number of available dual-radio nodes, analysis of deployments involving larger numbers of nodes relied on simulations.

The performance of IEEE 802.15.4 is investigated by simulating a very low bit-rate service, namely, an audio stream with a rate

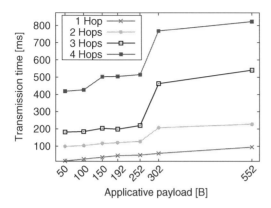

Figure 3.24 Experimental results: transmission time as a function of the payload size.

of 1008 Byte/s. This bit rate was chosen since Contiki has severe constraints on the dimension of the IPv6 buffer, which limits the size of the payload of IPv6 packets. Low-power devices, such as TelosB motes, have very limited RAM – of the order of a few kilobytes – so that memory allocation to build and send IPv6 packets is a practical issue.

In Figure 3.24, the transmission time is shown as a function of the payload. In particular, the results show that increasing the packet payload does not significantly affect the performance of the network for buffer-size dimensions below 300 B (i.e., 252 B at the application layer). On the contrary, for larger packet dimensions, the packet is fragmented, thus leading to higher transmission latency.

At this point, we set the packet size to 300 B and we configure the nodes to exchange data using UDP over IPv6. Thus, the packet is structured as follows:

- 40 B of IPv6 header
- 8 B of UDP header
- 252 B of application data.

With this configuration, we investigate the feasibility of the IEEE 802.15.4 standard for media streaming. In particular, we simulate in Cooja the transmission of a 5 s audio stream (at 1008 B/s) in a multi-hop network. Even this low rate can have a negative impact on network performance, in terms of packet-loss rate, as shown in Table 3.7. In fact, the packet loss reaches 45% with six hops and 50%

Table 3.7 Packet-loss rate at different hops for a 5 s audio stream at 1008 B/s (i.e., 5 kB).

Number of hops	1	2	3	4	6	8	10
Packet-loss rate	0%	5%	5%	15%	45%	50%	50%

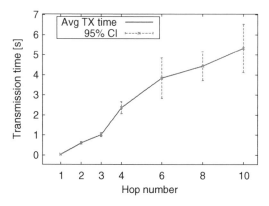

Figure 3.25 Experimental results: average transmission time with 95% Confidence Interval (CI), as a function of the number of hops from source to destination.

with eight or more hops. The packet loss under the IEEE 802.15.4 protocol is mainly due to collisions between packets and limited buffer queues. In fact, when a node acting as relay tries to forward a packet, it may still find the medium busy because of another data transmission. Since queue sizes at nodes are limited and intermediate nodes cannot store all the received data, the network experiences significant packet loss.

Figure 3.25 shows the average time required to send an IPv6 packet – part of a 20-packet stream – between source and destination, as a function of the number of hops. As expected, this time is approximately a linearly increasing function of the number of hops.

In Figure 3.26, the performance of an IEEE 802.15.4 network, in terms of packet-loss rate, is shown as a function of the number of hops. In this case, the offered load on the network is given by a 1 min long 1008 B/s stream; that is, 60kB. This makes the performance even worse: the packet-loss rate increases to 39% for a four-hop transmission and to 53.75% for a six-hop transmission.

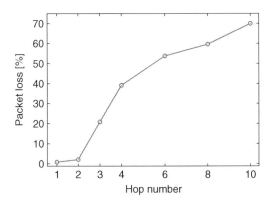

Figure 3.26 Experimental results: packet-loss rate as a function of the number of hops from source to destination.

The results show that a multi-hop IEEE 802.15.4 network cannot meet the strict requirements of real-time streaming applications, in terms of packet-loss rate and delay, in networks where communication routes may have more than four hops. However, in the case of very low-rate audio streaming over networks with a small value of the maximum number of hops, such requirements could be met. It must be remarked that there will always be a limitation on the stream duration: injecting more data into the network (say, by transmitting for 1 min at 1008 B/s) may lead to a breakdown of the IEEE 802.15.4 network, thus making communications impractical. Therefore, high-bandwidth and high-performance services cannot run on IEEE 802.15.4 networks.

3.8.6 IEEE 802.15.4-controlled Selective Activation of the IEEE 802.11s Network

While IEEE 802.15.4 is suitable for lightweight tasks, requiring the transmission of only small amounts of data, as shown above it is not possible to rely on it for real-time, high-bandwidth, and low-latency applications. The next phase of our experimentation aims at determining the threshold amount of data that makes it more convenient to rely on DNMP to activate a HPHT network. We refer to the same linear topology considered in Section 3.8.5.1: while for the IEEE 802.15.4 network, four hops are needed, only two hops are needed for IEEE 802.11s network. We consider the transmission of a fixed amount of application data (payload) and analyze the time required to transmit the payload with the proposed hybrid IEEE 802.15.4/IEEE 802.11s approach (with DNMP). As considered in Section 3.8.5.2, in the IEEE 802.15.4 network the payload is inserted into UDP/6LoWPAN packets of 300 B (i.e., 252 B of payload per packet at the application layer). In the IEEE 802.11s network, the payload is sent in a standard UDP/IPV4 packet.

Referring to DNMP, the total transmission time is made up of:

- the transmission time of the DNMP LOCK message from the source to the destination (IEEE 802.15.4 network);
- the time needed to turn the IEEE 802.11s interfaces on (IEEE 802.11s network);
- the time to create high-speed routes successfully (IEEE 802.11s network).

The performance comparison was carried out by considering the mean round trip time (RTT) of a DNMP message in the deployed five-node IEEE 802.15.4 network. In particular, the RTT accounts for:

- the transmission of the DNMP request message;
- the processing of the message at the application layer by Contiki OS on each relay node;
- the transmission of the response back to the source node.

The time between the instant when the IEEE 802.11s interface is switched on and the instant when the transmission towards the destination is possible has been experimentally measured at 3.237 s. This delay is due to kernel calls that start the radio networking and the traffic needed by the HWMP protocol to set up the routing.

In Figure 3.27, the time (dimension: [s]) required for the data to reach the destination is shown as a function of the payload size (dimension: [B]). While the time required using only the IEEE 802.15.4 network increases significantly with the payload size, with DNMP, after an initial setup time, the transmission time still increases, but almost negligibly with respect to the single IEEE 802.15.4 network case. The results show that the threshold below which the DNMP-based approach becomes inefficient is ~2 kB.

3.8.7 Conclusions

DNMP is an innovative IP-based application-layer protocol that allows high-bandwidth radio interfaces to be adaptively turned

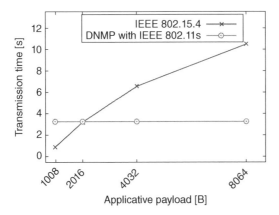

Figure 3.27 Experimental results: transmission time as a function of the payload size using IEEE 802.15.4 and IEEE 802.11s.

on and off, in order to support high-throughput and low-latency transmissions. In particular, it only requires the presence of an LPLT interface (acting as both data and control plane) to activate an HPHT interface (data plane). By leveraging on the routes created by the LPLT network, nodes manage to signal and wake up only the minimum set of nodes that guarantee data transmission via the HPHT network. In order to provide an exhaustive performance evaluation of DNMP, we integrated an IEEE 802.15.4 transceiver (LPLT) and an IEEE 802.11s (HPHT) transceiver on a Raspberry Pi. We then investigated the performance of the DNMP-based overlaid networks, determining the critical threshold, in terms of amount of data to be transmitted, beyond which the use of DNMP becomes advantageous. In the considered experimental scenario this critical threshold is ~2 kB.

3.9 Discoverability in Constrained Environments

3.9.1 CoRE Link Format

The Constrained RESTful Environments (CoRE) approach implements the REST architecture in way that supports constrained smart objects and networks (e.g., 6LoWPAN). The discovery of resources hosted by a server is fundamental for M2M scenarios. The HTTP "web discovery" of resources is defined in RFC 2616 [2] and the description of relations between resources is defined as "web linking" [12]. The CoRE Link Format specification [13] defines the discovery of resources hosted by a constrained web server, their attributes, and other resource relations such as CoRE resource discovery.

The main function of such a discovery mechanism is to provide universal resource identifiers (URIs, commonly called "links") for the resources hosted by the smart object. The resources are complemented by attributes about those resources and possible further link relations. In CoRE, this collection of links is represented as a resource of its own (in contrast to the HTTP headers, which are delivered along with a specific resource). The RFC document specifies a link format for use in CoRE resource discovery by extending the HTTP Link header format [12] to describe these link descriptions. The CoRE Link Format is carried as a payload and is assigned an Internet media type. A "well-known" relative URI /.well-known/ core is defined as a default entry point for requesting the list of links about resources hosted by a server node and thus performing CoRE resource discovery. This approach is suitable for adoption in combination with CoAP, HTTP, or any other web transfer protocol. The link format can also be saved in file format.

The CoRE Link Format is a serialization of a typed link describing the relationships between resources: so-called "web linking". In the RFC, web linking is extended with specific constrained M2M attributes; links are carried as a message payload rather than in an HTTP link header field, and a default interface is defined to discover resources hosted by a server. This specification also defines a new relation type: "hosts". These identify that the resource is hosted by the server from which the link document was requested.

In HTTP, the link header is used to keep link information about a resource along with an HTTP response. In CoRE, the main use case for web linking is the discovery of which resources a server hosts in the first place. Some resources may have additional links associated with them and, for that reason, the CoRE link format serialization is carried as a resource representation of the well-known URI. The CoRE link format reuses the format of HTTP link header serialization.

3.9.1.1   CoRE Link Format: Discovery

In IoT applications, there is the concrete need for local clients and servers to find and interact with each other without human intervention or any prior configuration. The CoRE Link Format can be used by smart objects hosting resources in such environments so as to enable discovery of the resources hosted by the server. Resource discovery can be performed either in unicast or multicast mode. When a server's IP address is already known, either a priori or resolved via the DNS [50, 51], unicast discovery is performed in order to locate the entry point to the resource of interest. In CoRE Link Format, this is performed using a GET to `/.well-known/core` on the server, which returns a payload in the CoRE Link Format. A client would then match the appropriate resource type, interface description, and possible media type [52] for its application. These attributes may also be included in the query string in order to filter the number of links returned in a response.

Multicast resource discovery is useful when a consumer wants to locate a resource within a limited local scope, and that scope supports IP multicast. A GET request to the appropriate multicast address is made for `/.well-known/core`. In order to limit the number and size of responses, a query string is submitted with the known attributes of the resource. Typically, a resource would be discovered based on its resource type and/or interface description, along with possible application-specific attributes.

3.9.1.2 Link Format

The CoRE Link Format (see Listing 3.1) extends the HTTP Link header field specified in RFC 5988 [12]. The format is compact and extensible does not require special XML or binary parsing. This link format is just one serialization of the typed links defined in the stardard; others include HTML links, Atom feed links [53], and HTTP Link header fields. The CoRE link format defines the Internet media type "application/link-format", encoded as UTF-8. UTF-8 data can be compared bitwise, which allows values to contain UTF-8 data without any added complexity for constrained nodes. The CoRE Link Format is equivalent to the RFC 5988 ([12]) Link Format, but the augmented Backus-Naur form in the present specification is repeated along with some improvements, so as to be compliant with RFC 5234 [54]. It also includes new link parameters. The link parameter "href" is reserved for use as a query parameter for filtering in this specification, and must not be defined as a link parameter. Multiple link descriptions are separated by commas. Note that commas can also occur in quoted strings and URIs but do not end a description. In order to convert an HTTP link header field to this link format, the "Link:" HTTP header is removed, any linear whitespace is removed, the header value is converted to UTF-8, and any percent- encodings are decoded.

Listing 3.1 CoRE link format specification as in RFC 6690.

```
Link             = link-value-list
  link-value-list = [ link-value *[ "," link-value ]]
  link-value      = "<" URI-Reference ">"
                    *( ";" link-param )
  link-param       = ( ( "rel" "=" relation-types )
                    / ( "anchor" "=" DQUOTE URI-Reference
                          DQUOTE )
                    / ( "rev" "=" relation-types )
                    / ( "hreflang" "=" Language-Tag )
                    / ( "media" "=" ( MediaDesc
                          / ( DQUOTE MediaDesc DQUOTE ) ) )
                    / ( "title" "=" quoted-string )
                    / ( "title*" "=" ext-value )
                    / ( "type" "=" ( media-type
                    / quoted-mt ) )
                    / ( "rt" "=" relation-types )
                    / ( "if" "=" relation-types )
                    / ( "sz" "=" cardinal )
                    / ( link-extension ) )
  link-extension = ( parmname [ "=" ( ptoken
    / quoted-string ) ] )
    / ( ext-name-star "=" ext-value )
  ext-name-star  = parmname "*" ; reserved
                                for RFC-2231-profiled
                              ; extensions.
                                Whitespace NOT
                              ; allowed in between.
  ptoken         = 1*ptokenchar
  ptokenchar     = "!" / "#" / "$" / "%" / "&" / "'" / "("
                  / ")" / "*" / "+" / "-" / "." / "/"
                  / DIGIT
                  / ":" / "<" / "=" / ">" / "?" / "@"
                  / ALPHA
                  / "[" / "]" / "^" / "_" / "'" / "{" / "|"
                  / "}" / "~"
  media-type     = type-name "/" subtype-name
  quoted-mt      = DQUOTE media-type DQUOTE
  relation-types = relation-type
                  / DQUOTE relation-type
                  *( 1*SP relation-type ) DQUOTE
  relation-type  = reg-rel-type / ext-rel-type
  reg-rel-type   = LOALPHA *( LOALPHA / DIGIT / "." / "-" )
  ext-rel-type   = URI
  cardinal       = "0" / ( %x31-39 *DIGIT )
  LOALPHA        = %x61-7A   ; a-z
  quoted-string  = <defined in [RFC2616]>
```

```
URI             = <defined in [RFC3986]>
URI-Reference   = <defined in [RFC3986]>
type-name       = <defined in [RFC4288]>
subtype-name    = <defined in [RFC4288]>
MediaDesc       = <defined in [W3C.HTML.4.01]>
Language-Tag    = <defined in [RFC5646]>
ext-value       = <defined in [RFC5987]>
parmname        = <defined in [RFC5987]>
```

3.9.1.3   The Interface Description Attribute

The interface description "if" attribute is an opaque string used to provide a name or URI indicating a specific interface definition used to interact with the target resource. This attribute describes the generic REST interface to interact with a resource or a set of resources. It is expected that an interface description will be reused by different resource types, for example the resource types "outdoor-temperature", "dew-point", and "rel-humidity". Multiple interface descriptions *may* be included in the value of this parameter, each separated by a space, similar to the relation attribute.

The interface description can be also the URI of a web application description language (WADL) definition of the target resource `http://www.example.org/myapp.wadl#sensor`, a URN indicating the type of interface to the resource, such as "urn:myapp:sensor", or an application-specific name, such as "sensor". The Interface Description attribute must not appear more than once in a link.

3.9.2   CoRE Interfaces

The resource discovery offered by a smart object is a fundamental element in IoT applications. The discovery of resources provided by an HTTP web server is defined by the web linking RFC [12] and its adoption for the description and discovery of resources hosted by smart objects is specified by the CoRE Link Format 3.9.1 and can be used by CoAP or HTTP servers. The CoRE Link Format defines an attribute that can be used to describe the REST interface of a resource, and may include a link to a description document.

The *CoRE Interfaces* document [55] defines the well-known REST interface descriptions for resource types in constrained environments using the CoRE Link Format standard. A short reference is provided for each type that can be efficiently included in the interface description attribute (if=) of the CoRE Link Format. A client discovering the "if" link attribute will be able to consume resources based on its knowledge of the expected interface types. In this sense, the interface type acts in a similar way as a content-format, but as a selector for a high-level functional abstraction.

The main defined REST interfaces are related to the following resources:

- sensor
- parameter
- read-only parameter
- actuator.

Each of them is described with the corresponding value of the interface description attribute (if) and the associated valid methods. These interfaces can support plain text and/or Sensor Markup Language (SenML) media types (see Section 3.10). Table 3.8 and Sections 3.9.2.1–3.9.2.4 will present available interfaces and their characteristics.

When a value for the if= attribute appears in a link, the associated resource must implement and support the associated standard REST interface and may support additional functionality. This approach has been designed to work within the context of CoRE link format specifications, but is applicable for REST interface definitions.

3.9.2.1 Sensor

The sensor interface allows the reader to access the value of a sensor resource through a GET request. The media type of the resource can be plain text or SenML. The former may be used for a single measurement that does not require metadata and, for a measurement with metadata such as a unit or time stamp, SenML is the best approach. A resource with this type of interface can also use SenML to

Table 3.8 Interface description summary.

Interface	if=	Methods	Content-Formats
Sensor	core.s	GET	link-format,text/plain
Parameter	core.p	GET, PUT	link-format,text/plain
Read-only Parameter	core.rp	GET	link-format,text/plain
Actuator	core.a	GET, PUT, POST	link-format,text/plain

return multiple measurements in the same representation. Listing 3.2 reports some examples of sensor interface requests in both text/plain and application/senml+json formats.

Listing 3.2 Examples of sensor interface requests.

```
Req: GET /s/humidity (Accept: text/plain)
  Res: 2.05 Content (text/plain)
  80
  Req: GET /s/humidity (Accept: application/senml+json)
  Res: 2.05 Content (application/senml+json)
  {"e":[
      { "n": "humidity", "v": 80, "u": "%RH" }],
  }
```

3.9.2.2 Parameter

The parameter interface allows configurable parameters/information to be modeled as a resource. The value of the parameter can be read using a GET or updated using a PUT request. Both plain text or SenML media types can be returned. The following example shows a request for reading and updating a parameter resource.

Listing 3.3 Request for reading and updating a parameter resource.

```
Req: GET /d/name
  Res: 2.05 Content (text/plain)
  node5
  Req: PUT /d/name (text/plain)
  outdoor
  Res: 2.04 Changed
```

3.9.2.3 Read-only Parameter

Following the same approach as for the parameter interface, the read-only parameter interface only allows the reader to access the parameter's configuration using a GET request. Plain text or SenML media types may be returned from this type of interface and Listing 3.4 shows an example.

Listing 3.4 Example of read-only parameter interface.

```
Req: GET /d/model
  Res: 2.05 Content (text/plain)
  SuperNode200
```

3.9.2.4 Actuator

The actuator interface is associated with resources that model different kinds of actuators. Examples of actuators include LEDs, relays, motor controllers and light dimmers. The value associated to the the actuator can be read using a GET and updated through a PUT request. Furthermore, this interface allows the use of POST to change the state of an actuator, for example to toggle between its possible values. Plain text or SenML media types can be used and returned for this interface category. SenML can be also used to include multiple measurements in the same representation. An example would be a list of recent actuator values or a list of values to update. Listing 3.5 shows requests for reading, setting and toggling an actuator (associated with an LED).

Listing 3.5 Requests to change an actuator.

```
Req: GET /a/1/led
Res: 2.05 Content (text/plain)
0
Req: PUT /a/1/led (text/plain)
1
Res: 2.04 Changed
Req: POST /a/1/led (text/plain)
Res: 2.04 Changed
Req: GET /a/1/led
Res: 2.05 Content (text/plain)
0
```

3.10 Data Formats: Media Types for Sensor Markup Language

Connecting sensors to the internet is not new, and there have been many protocols designed to facilitate the process. The SenML specification [56] defines new media types to embed simple sensor information in application protocols, such as HTTP and CoAP. The format was designed so that processors with very limited capabilities could easily encode a sensor measurement into the media type, while at the same time a server parsing the data could relatively efficiently collect a large number of sensor measurements. SenML is defined by a data model for measurements and simple metadata about measurements and devices. The data is structured as a single object (with attributes) that contains an array of entries. Each entry is an object that has attributes, such as a unique identifier for the sensor, the time the measurement was made, and the current value. Serializations for this data model are defined for JSON, XML and Efficient XML Interchange (EXI).

This approach allows a server to efficiently parse large numbers of measurements. SenML can be adopted to data flow models, data feeds pushed from a sensor to a collector, and to the web resource model, where the sensor is requested as a resource representation (e.g., "GET/sensor/temperature").

SenML strikes a balance between having some information about the sensor carried with the sensor data so that the data is self-describing but also making that a fairly minimal set of auxiliary information, for efficiency reasons. Other information about the sensor can be discovered by other methods such as using the CoRE link format.

For example, Listing 3.6 shows a measurement from a temperature gauge encoded in the JSON syntax. The array has a single SenML record with a measurement for a sensor labelled as "urn:dev:ow:10e2073a01080063" with a current value of 23.1° C.

Listing 3.6 Temperature measurement encoded in JSON.

```
[
    {"n":"urn:dev:ow:10e2073a01080063","u":"Cel","v":23.1}
]
```

SenML is defined by a data model for measurements and simple metadata about measurements and devices. The data is structured as a single array containing a list of SenML records. Each record contains fields such as an unique identifier for the sensor, the time, the measurement, the unit the measurement is recorded in, and the current value of the sensor. Serializations for this data model are defined for JSON, CBOR, XML, and Efficient XML Interchange (EXI). SenML's main concepts and elements are:

- *SenML record:* One measurement or configuration instance in time presented using the SenML data model.
- *SenML pack:* One or more SenML records in an array structure.
- *SenML label:* A short name used in SenML records to denote different SenML fields (e.g., "v" for "value").
- *SenML field:* A component of a record that associates a value to a SenML label for this record.

Each SenML pack carries a single array that represents a set of measurements and/or parameters. This array contains a series of SenML records with various fields, as described below. There are two kinds of fields: base and regular. The base fields can be included in any SenML record and they apply to the entries in the record. Each base field also applies to all records after it, up to, but not including the next record that has that same base field. All base fields are optional. Regular fields can be included in any SenML record and apply only to that record. Basic fields associated with SenML records are:

- *Base name:* A string prepended to the names found in the entries.
- *Base time:* A base time that is added to the time found in an entry.
- *Base unit:* A base unit that is assumed for all entries, unless otherwise indicated. If a record does not contain a unit value, then the base unit is used. Otherwise the value found in the unit (if any) is used.
- *Base Value:* A base value is added to the value found in an entry, similar to base time.
- *Base sum:* A base sum is added to the sum found in an entry, similar to base time.
- *Version:* Version number of media type format. This field is an optional positive integer and defaults to 5 if not present.

Regular fields are:

- *Name:* Name of the sensor or parameter. When appended to the base n Name field, this must result in a globally unique identifier for the resource. The name is optional if the base name is present. If the name is missing, base name must uniquely identify the resource. This can be used to represent a large array of measurements from the same sensor without having to repeat its identifier on every measurement.
- *Unit:* Units for a measurement value. Optional.
- *Value:* Value of the entry. Optional if a sum value is present, otherwise required. Values are represented using basic data types. This specification defines floating point numbers ("v" field for value), Booleans ("vb" for Boolean value), strings ("vs" for string value) and binary data ("vd" for data value). Exactly one value field *must* appear unless there is sum field, in which case it is allowed to have no value field.
- *Sum:* Integrated sum of the values over time. Optional. This field is in the units specified in the unit value multiplied by seconds.
- *Time:* Time when value was recorded. Optional.

Table 3.9 JSON SenML labels.

SenML	Label JSON	Type
Base name	bn	String
Base time	bt	Number
Base units	bu	String
Base balue	bv	Number
Base dum	bs	Number
Version	bver	Number
Name	n	String
Unit	u	String
Value	v	Number
String value	vs	String
Boolean value	vb	Boolean
Data value	vd	String
Value sum	s	Number
Time	t	Number
Update time	ut	Number
Link	l	String

- *Update time:* An optional time in seconds that represents the maximum time before this sensor will provide an updated reading for a measurement. This can be used to detect the failure of sensors or communications paths from them.

Table 3.9 shows the JSON representation (application/senml+json) for SenML fields. The SenML labels are used as the JSON object member names in JSON objects representing JSON SenML records. Examples of SenML use are presented in Section 3.10.1.

3.10.1   JSON Representations

3.10.1.1   Single Datapoint

Listing 3.7 shows a temperature reading taken approximately "now" by a one-wire sensor device that was assigned the unique one-wire address of 10e2073a01080063.

Listing 3.7 Temperature reading from one-wire device at unspecified time.

```
[
  {"n":"urn:dev:ow:10e2073a01080063","u":"Cel","v":23.1}
]
```

3.10.1.2 Multiple Datapoints

Listing 3.8 shows voltage and current "now" in the JSON representation; that is at an unspecified time.

Listing 3.8 Voltage and current measurements at unspecified time.

```
[
  {"bn":"urn:dev:ow:10e2073a01080063:","n":"voltage",
   "u":"V","v":120.1},
  {"n":"current","u":"A","v":1.2}
]
```

Listing 3.9 is similar, but shows the current on Tuesday 9 June 2010 at 18:01:16.001 UTC and at each second for the previous 5 s.

Listing 3.9 Temperature readings at specified dates and times.

```
[
  {"bn":"urn:dev:ow:10e2073a0108006:",
   "bt":1.276020076001e+09,"bu":"A","bver":5,
   "n":"voltage","u":"V","v":120.1},
  {"n":"current","t":-5,"v":1.2},
  {"n":"current","t":-4,"v":1.3},
  {"n":"current","t":-3,"v":1.4},
  {"n":"current","t":-2,"v":1.5},
  {"n":"current","t":-1,"v":1.6},
  {"n":"current","v":1.7}
]
```

3.10.1.3 Multiple Measurements

Listing 3.10 shows humidity measurements from a mobile device with a one-wire address 10e2073a01080063, starting at Monday 31 October 2011 at 13:24:24 UTC. The device also provides position data, which is provided in the same measurement or parameter array as separate entries. Note that the time is used for correlating data that belongs together, such as a measurement and a parameter associated with it. Finally, the device also reports extra data about its battery status at a separate time.

Listing 3.10 Multiple humidity measurements.

```
[
  {"bn":"urn:dev:ow:10e2073a01080063",
   "bt":1.320067464e+09,"bu":"%RH","v":20},
  {"u":"lon","v":24.30621},
  {"u":"lat","v":60.07965},
  {"t":60,"v":20.3},
  {"u":"lon","t":60,"v":24.30622},
  {"u":"lat","t":60,"v":60.07965},
  {"t":120,"v":20.7},
  {"u":"lon","t":120,"v":24.30623},
  {"u":"lat","t":120,"v":60.07966},
  {"u":"%EL","t":150,"v":98},
  {"t":180,"v":21.2},
  {"u":"lon","t":180,"v":24.30628},
  {"u":"lat","t":180,"v":60.07967}
]
```

4

Discoverability

The IoT is envisioned to bring together billions of devices, or "smart objects", by connecting them in an Internet-like structure, allowing them to communicate and exchange information and to enable new forms of interaction among things and people. Smart objects are typically equipped with a microcontroller, a radio interface for communication, sensors and/or actuators. Smart objects are constrained devices, with limited capabilities in terms of computational power and memory. They are typically battery-powered, thus introducing even more constraints on energy consumption: this motivates the quest for energy-efficient technologies, communication/networking protocols, and mechanisms. Internet Protocol (IP) has been widely envisaged as the true IoT enabler, as it allows full interoperability among heterogeneous objects. As part of the standardization process that is taking place, new low-power protocols are being defined in international organizations, such as the IETF and the IEEE.

4.1   Service and Resource Discovery

Together with application-layer protocols, suitable mechanisms for service and resource discovery should be defined. In particular, CoAP defines the term *service discovery* as the procedure used by a client to learn about the endpoints exposed by a server. A service is discovered by a client by learning the Uniform Resource Identifier (URI) [28] that references a resource in the server namespace. *Resource discovery* is related to the discovery of the resources offered by a CoAP endpoint. In particular, M2M applications rely on this feature to keep applications resilient to change, and therefore not requiring human

Internet of Things: Architectures, Protocols and Standards, First Edition.
Simone Cirani, Gianluigi Ferrari, Marco Picone, and Luca Veltri.
© 2019 John Wiley & Sons Ltd. Published 2019 by John Wiley & Sons Ltd.

intervention. A *resource directory* (RD) [14] is a network element hosting the description of resources held on other servers, allowing lookups to be performed for those resources.

A crucial issue for the robust applications, in terms of resilience to changes that might occur over time (e.g., availability, mobility, or resource description), and the feasible deployment of (billions of) smart objects is the availability of mechanisms that minimize, if not remove, the need for human intervention for the configuration of newly deployed objects. The RESTful paradigm is intended to promote software longevity and independent evolution [29], both of which are extremely important for IoT and M2M applications deployed on smart objects that are expected to stay operational for long periods; say, years. Self-configuring service and resource discovery mechanisms should take into account the different scopes that these operations might have:

- in a local scope, they should enable communication between geographically concentrated smart objects; that is, residing in the same network;
- in a global (large-scale) scope, they should enable communication between smart objects residing in different (and perhaps geographically distant) networks.

These approaches should also be scalable, since the expected number of deployed objects is going to be of the order of billions.

Self-configuration is another crucial feature for the diffusion of IoT systems, where all the objects equipped with a radio interface are potential sources of information to be interconnected. An external operator managing a network first needs to configure the system. Clearly, if this operation is carried out manually, there may be misconfigurations. This is far more likely when thousands of devices are involved. In addition, an occasional manual network reconfiguration may cause a significant system outage, just as, in an industrial plant, machines may need to be stopped for normal maintenance. For this reason, a self-configurable IoT system is a good way to prevent long outages and configuration errors.

4.2   Local and Large-scale Service Discovery

In the literature, there are already mechanisms for implementing service discovery. Most of these, however, were originally conceived for

LANs and were then extended for constrained IPv6 over low-power wireless personal area networks (6LoWPANs). One of these mechanisms is Universal Plug and Play (UPnP) [57], a protocol that allows for automatic creation of a device-to-device network. However, as UPnP uses TCP as the transport protocol and XML as the message exchange format, it is not suited to constrained devices.

Another proposed mechanism is based on the Service Location Protocol (SLP) [58, 59] through which computers and devices can find services in LANs without prior configuration. Devices use SLP to announce the services they provide in the local network; these are grouped into scopes: simple strings that classify the services. The use of SLP may be important in large-scale IoT scenarios, in order to make service discovery automatic. However, SLP does not target constrained devices like those used in the IoT. In addition, it relies on centralized approaches, which may be prone to failure. Finally, up to now, no SLP implementation has been available for Contiki-based devices.

Another alternative to UPnP is the Zero-configuration (Zero-conf) [60] networking protocol, which allows for automatic creation of computer networks based on the TCP/IP Internet stack and does not require any external configuration. Zeroconf implements three main functionalities:

- automatic network address assignment;
- automatic distribution and resolution of host names;
- automatic location of network services.

Automatic network assignment intervenes when a node first connects to the network. The host name distribution and resolution is implemented using multicast DNS (mDNS) [61], a service that has the same interfaces, packet formats, and semantics as standard DNS messages to resolve host names in networks that do not include a local name server. In the service discovery phase, Zeroconf implements DNS-based Service Discovery (DNS-SD) [62]. Using standard DNS queries, a client can discover, for a given domain, the named instances of the service of interest.

In the field of ubiquitous computing, PIAX, a P2P platform for geographic service location, has been proposed [63, 64]. In PIAX, every node is a peer of the overlay. This approach is not suitable for the IoT, since many nodes are constrained in terms of processing capabilities. In addition, PIAX does not provide a URI resolution service, so that it can only try to route the query to the correct area of the network but cannot resolve the endpoint to be contacted.

Efforts have been made to adapt these solutions to the world of constrained devices. Busnel *et al.* introduced a P2P overlay to perform broadcast or anycast in wireless sensor networks (WSNs) without any centralized element [65]. Sensors were clustered according to their types into specific layers. However, they took into account neither local service discovery nor computational complexity due to the existence of nodes belonging to different layers.

Gutierrez *et al.* instead introduced a separation between WSNs and P2P networks [66]. Their focus was on exploiting these two types of network to develop a feedback loop to allow developers to define self-managing behaviors. However, they did not take into account aspects like energy efficiency, self-discovery of resources, or large-scale deployments.

Leguay *et al.* implemented an automatic discovery mechanism [67]. In their approach each node is responsible for announcing itself to the main gateway through HELLO messages. These messages are sent either in response to a discovery request or proactively sent in an automatic way. The gateway is then be in charge of addressing the requests coming from external networks to the correct nodes.

Kovacevic *et al.* have proposed NanoSD, a lightweight service discovery protocol designed for highly dynamic, mobile, and heterogeneous sensor networks [68]. Their solution requires extensive multicast and broadcast messages to keep track of service information of the neighboring nodes.

Another solution was presented by Mayer and Guinard [69]. They developed a RESTFul web service using HTTP-based service discovery. However, their approach does not provide management and status maintenance of existing services.

Finally, Butt *et al.* divided the network into groups, assigning different roles to the nodes in each group [70]. Embedding a directory agent into the border router makes scalability easier. However, this architecture tends to be too fragile in the presence of failures of the central border router. In addition, the protocol focuses on in-network service location, but it lacks coordination with other similar entities, thus preventing large-scale discovery.

A few papers related to service discovery in IoT systems have appeared. Jara *et al.* sketched an architecture for large-scale service discovery and location [71]. However, theirs was a centralized solution, exposing a search engine to make the integration of distributed service directories feasible.

Paganelli and Parlanti exploited an underlying distributed P2P overlay to support more complex queries, such as multi-attribute and range queries [72]. This approach is more focused on service resolution rather than on the creation of the overlay by automatically discovering existing services. Unlike our approach, which aims at being transparent and agnostic of the underlying technology, several P2P overlays presented in the literature focus on RFID for supply chains [73–75].

CoAP natively provides a mechanism for service discovery and location [7]. Each CoAP server must expose an interface /.well-known/core, to which the RD or, more generally, a generic node can send requests for discovering available resources. The CoAP server will reply with the list of resources and, for each resource, an attribute that specifies the format of the data associated with that resource. CoAP, however, does not specify how a node joining the network for the first time must behave in order to announce itself to the resource directory node.

In the IETF's latest draft for CoAP [76], this functionality is extended to multicast communications. In particular, multicast resource discovery can be useful when a client needs to locate a resource within a limited scope, and that scope supports IP multicast. A GET request to the appropriate multicast address is made for /.well-known/core. Of course, multicast resource Discovery works only within an IP multicast domain and does not scale to larger networks that do not support end-to-end multicast. However, in CoAP there is no specification on how a remote client can lookup the RD and query for the resource of interest.

Peer-to-peer (P2P) networks have been designed to provide some desirable features for large-scale systems, such as scalability, fault-tolerance, and self-configuration. The main feature that makes P2P networks appealing is the fact that as the number of participating nodes increases, the overall system capacity (in terms of processing and storage capabilities) increases as well. This challenges classical client/server architectures, where an increase in the number of clients may bring the system to saturation and/or failure. P2P networks arrange participating nodes in an overlay network, built on top of an existing network, such as the Internet. The algorithm through which the overlay is created can be used to make a distinction between structured and unstructured P2P networks. Structured P2P networks, such as distributed hash tables (DHTs), are built using *consistent hashing* algorithms, which guarantee that the routing of requests takes a deterministic and upper-bounded number of hops for completion, at the cost of having network traffic for managing and maintaining the overlay. Historically, P2P networks have been associated with file sharing applications, such as eMule[1] and BitTorrent[2]. The decrease in the popularity of file sharing applications has cooled interest in P2P, even though notable applications, such as Skype, have historically used a P2P overlay as backbone to provide a scalable and efficient service. However, the features that P2P networks have been designed for are very appealing for IoT scenarios, where large-scale and robust applications need to be supported. IoT thus represents an opportunity of a renaissance for P2P.

Centralized approaches for service discovery, such as the RD of the CoAP protocol, suffer from scalability and availability limitations and are prone to attacks, such as *denial of service* (DoS). Possible alternatives to this problem may consist of the use of DHTs. Key/value pairs are stored in a DHT and any participating node can efficiently retrieve the value associated with a given key. Responsibility for maintaining the mapping from keys to values is distributed among the nodes in such a way that a change in the set of participants causes a minimal amount of disruption (consistent hashing). This allows a DHT to scale to extremely large numbers of nodes and to handle continuous node arrivals, departures, and failures.

Several different algorithms and protocols have already been proposed for DHTs; the most significant are Chord [77] (for its simplicity) and Kademlia [78] (for its efficiency). Some papers have also been published on the use of P2P for service discovery. Yulin *et al.* combine P2P technology and the centralized Universal Description Discovery and Integration (UDDI) technology to provide a flexible and reliable service discovery approach [79]. Kaffille *et al.* apply the concepts of DHTs to the service discovery, creating an overlay P2P to exchange information about available services without flooding the entire network [80]. However, these approaches do not take into account the constraints and requirements of IoT. In Section 4.3, we will detail our P2P implementation for large-scale service/resource discovery in IoT networks, extending the P2P DHT solution by taking

---

1 http://www.emule-project.net/.
2 http://www.bittorrent.com/.

into account the requirements of scalability and self-configuration typical of constrained networks.

4.2.1 ZeroConf

ZeroConf is an open standard originally designed by Apple. It allows services to be setup automatically within a network, without requiring manual configuration. The IETF Zeroconf Working Group was formed in 1999 and has worked on the definitions and the standardization mechanisms required to achieve zero configuration of services.

ZeroConf is based on the combination of three functionalities: address selection, service name resolution, and service discovery. These functions are provided by the following suite of standards, respectively:

- IPv4 link-local addressing [81]: this standard allows hosts to self-assign IP addresses in a network without relying on a DHCP server;
- Multicast DNS [82]: this standard provides a way to resolve names to IP addresses without relying on a DNS server;
- DNS service discovery (DNS-SD) [83]: this standard allows discovery of services within a network using the semantics of DNS messages.

The ZeroConf suite allows services to be configured and discovered without requiring DHCP and DNS servers. It does this by making all hosts collectively responsible for publishing/discovering services and resolving names to addresses, simply by using the semantics of the DNS protocol and multicast communication.

ZeroConf supports both IPv4 and IPv6, using multicast IP addresses 224.0.0.251 and ff02::fb and UDP ports 53 and 5353, respectively.

Service discovery typically occurs by searching for services of a given type in a particular domain; that is, matching the service type string of the form _ServiceName._ServiceType._TransportProtocolName. Domain. (for example, _http._tcp.local. or _coap._udp.local.) The service discovery process returns a list of services that comply with the searched type. Subsequently, a service with a given name can be resolved to an IP address and port number at which it can be accessed. In the case of a CoAP server, once discovered, a request can be issued to the server for its /.well-known/core in order to perform resource discovery.

Several implementations of ZeroConf are available, in essence for all platforms. Due to this widespread support, ZeroConf is a very good option for IoT smart objects that want to advertise their presence and allow other applications to discover them automatically with no manual configuration.

The limitation of ZeroConf is its reliance on multicast communication, which is rarely propagated beyond the scope of the local network, making ZeroConf typically suitable only for local environments. However, locally, ZeroConf provides an extremely convenient and elegant way to perform service discovery and can thus be adopted to deploy self-configuring IoT applications.

4.2.2   UPnP

With a similar intent to ZeroConf, the Universal Plug and Play (UPnP) protocol suite provides a way to perform dynamic and seamless discovery of devices in a network, without relying on DHCP and DNS servers. UPnP has been defined by the UPnP Forum and uses HTTPU/HTTPMU (HTTP over UDP and HTTP over multicast UDP) and SOAP to perform service description, discovery, and data transfer. UPnP suits home appliances rather than enterprise-level deployments due to its security and efficiency issues. Many consumer-oriented smart objects, such as Philips Hue light bulbs, use UPnP as a zero-configuration service discovery mechanism for bridges.

4.2.3   URI Beacons and the Physical Web

The Physical Web, a concept promoted by Google, is a different approach to provide seamless discovery and interaction with smart objects. The assumption behind the Physical Web is that the web itself provides all the necessary means for a fruitful interaction with any endpoint, be that a website or an object. As a consequence, the only operation that is needed in order to merge the physical world and the web is to discover the URL related to a web resource linked to a smart object. After that, a web browser is capable of delivering a user interface to the end user, which they can use to interact with the object (mediated by a backend that is actually connected to the object itself).

The discovery mechanism defined by the Physical Web (and illustrated in Figure 4.1) is based on the use of URI beacons; that is,

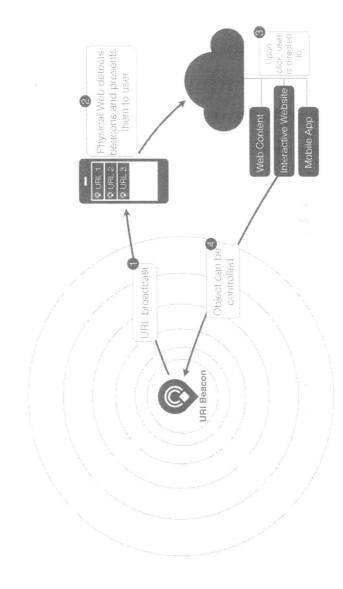

Figure 4.1 Physical Web discovery mechanism.

Physical Web detects beacons and presents them to user

Upon click, user is directed to

Web Content

Interactive Website

Mobile App

URL broadcast

Object can be controlled

URI Beacon

URL 1
URL 2
URL 3

Bluetooth Low Energy (BLE) devices broadcasting URLs. The use of BLE is particularly convenient because it is supported on the vast majority of user devices as well as having low energy consumption, which is important in order to ensure that battery-powered beacons can last as long as possible. The standard for data broadcasting over BLE is the Eddystone protocol, designed by Google. The Eddystone protocol defines four packet types:

- *Eddystone-UID,* used to contain a beacon identifier;
- *Eddystone-URL,* used to broadcast a URL;
- *Eddystone-TLM,* used for sending telemetry information
- *Eddystone-EID,* used for carrying ephemeral IDs, in order to protect against replay attacks or spoofing.

Although BLE is currently the only communication protocol that can be used to broadcast a URL, other options, such as mDNS or UPnP, can still be applied and might be supported in the future.

The advantage in using URI beacons is the possibility to discover and interact with objects even if the user device is not connected to the same network. However, this benefit may also become a downside, because the interaction with the object might not take into account context information related to the association of the user device with the network. Moreover, it may be unsafe in some scenarios to openly broadcast object URLs: it might raise security issues and it could be impossible to restrict discovery to only authorized devices. The Physical Web is therefore particularly suited to public spaces, where no restricted access to objects should occur.

4.3 Scalable and Self-configuring Architecture for Service Discovery in the IoT

In this section, we present a scalable and self-configuring architecture for service and resource discovery in the IoT. aiming at providing mechanisms requiring no human intervention for configuration, thus simplifying the deployment of IoT applications. Our approach is based on:

- at a large scale, P2P technologies, to provide a distributed large-scale service discovery infrastructure;
- at a local scale, zero-configuration mechanisms.

Information on resources provided by smart objects attached to a local wireless network are gathered by a special boundary node, referred to as the "IoT gateway". This is also part of a P2P overlay used to store and retrieve such information, resulting into a distributed and scalable RD. As will be shown, the global service discovery performance depends only on the number of peers in the P2P overlay; this makes the proposed approach directly scalable when the size of the IoT network increases. Local service discovery at the IoT gateway makes the process of discovery of new resources automatic. In particular, in our experimental tests we use CoAP for the description of the available endpoints.

To the best of our knowledge, this is the first research to provide an architecture and mechanisms that allow for service discovery at both global and local scales into a unique self-configuring system. We also provide some preliminary results obtained by an implementation and a real-world deployment of our architecture, thus demonstrating its feasibility.

We note that the proposed architecture is built upon components designed to be absolutely agnostic regarding the format of service and resource descriptors, in order to avoid the introduction of application-specific constraints. In fact, the architecture provides mechanisms for publishing and retrieving information, mapped to service or RD URIs, which can be represented in any suitable content format for service/resource description, either already available, such as the CoRE Link Format [13], or foreseeable. The adoption of standard description formats is mandatory to guarantee maximum interoperability, but it is a service's responsibility to enforce this practice. It is also important to note that IoT applications should be implemented according to the REST paradigm; the definition of CoAP is intended to accomplish precisely this. Client applications, in order to comply with the RESTful paradigm, must follow the HATEOAS (Hypermedia as the Engine of Application State) principle [2], which forbids applications from driving interactions that are based on out-of-band information rather than on hypermedia. The existence of prerequisites, in terms of resource representations, is a violation of the REST paradigm. The service discovery architecture itself does not do this: it is extremely flexible, able to handle any resource description format. The absence of content-related dependencies leads to more robust implementations, in terms of longevity and adaptability to changes that resource descriptions might undergo.

4.3.1 IoT Gateway

The service discovery architecture proposed in this work relies on the presence of an IoT gateway. By combining different functions, the IoT gateway provides both IoT nodes and standard (non-constrained) nodes with service and resource discovery, proxying, and (optionally) caching and access-control functionalities. In this section, the internal architecture of the IoT gateway and its associated functions will be detailed.

4.3.1.1 Proxy Functionality

The IoT gateway interacts, at the application level, with other IoT nodes through CoAP and may act as both CoAP client and CoAP server. More precisely, in the terms of the CoAP specifications, it may act as CoAP *origin server* and/or *proxy*. The CoAP specification defines an origin server as a CoAP server on which a given resource resides or has to be created, while a proxy is a CoAP endpoint which, by implementing both the server and client sides of CoAP, forwards requests to an origin server and relays back the received responses. The proxy may also (optionally) perform caching and protocol translation (in which case it is termed a "cross-proxy").

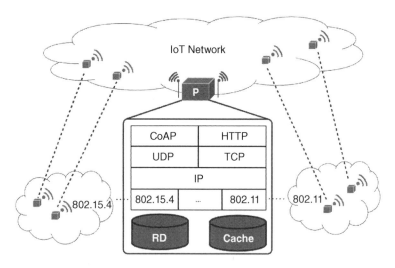

Figure 4.2 Architecture of IoT gateway with internal layers and caching/resource directory capabilities.

The presence of a proxy at the border of an IoT network can be very useful for a number of reasons:

- to protect the constrained network from the outside: for security reasons such as DoS attacks;
- to integrate with the existing web through legacy HTTP clients;
- to ensure high availability of resources through caching;
- to reduce network load of constrained devices;
- to support data formats that might not be suitable for constrained applications, such as XML.

In Figure 4.2, a layered view of the IoT gateway node is presented.

In addition to standard CoAP proxying behavior, the IoT gateway may also act as an HTTP-to-CoAP proxy by translating HTTP requests to CoAP requests (and vice-versa). Just like standard CoAP proxying, an HTTP-to-CoAP proxy can integrate two different operational modes:

- *reverse proxy*: by translating incoming HTTP requests to CoAP requests, it provides access to resources that are created and stored by CoAP nodes within the IoT network (acting as CoAP servers);
- *origin server*: acting as both HTTP and CoAP server, by letting CoAP nodes residing in the IoT network (and acting as clients) create resources through CoAP POST/PUT requests, and by making such resources available to other nodes through HTTP and CoAP.

The latter operational mode is particularly suited for duty-cycled IoT nodes, which may post resources only during short wake-up intervals. Figure 4.3 shows the difference between a reverse proxy and an origin server.

From an architectural point of view, the IoT gateway comprises the following elements:

- an IP gateway, managing IPv4/IPv6 connectivity among smart objects in heterogeneous networks (i.e., IEEE 802.15.4, IEEE 802.11.x, and IEEE 802.3) so as to allow for interconnection of devices operating in different networks by providing an IP layer to let nodes communicate seamlessly;
- a CoAP origin server, which can be used by CoAP clients within the network to post resources that will be maintained by the server on their behalf;

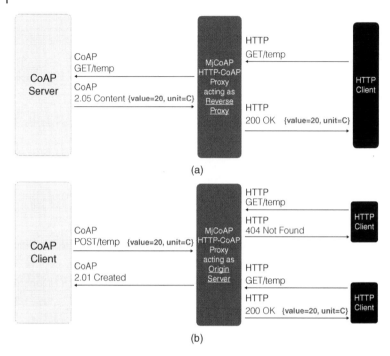

Figure 4.3 HTTP-to-CoAP proxy acting as: (a) reverse proxy and (b) origin server.

- a HTTP-to-CoAP reverse proxy, optionally equipped with caching capabilities, which can be used for accessing services and resources that are available in an internal constrained network.

The IoT gateway is therefore a network element that coordinates and enables full and seamless interoperability among highly heterogeneous devices, which:

- may operate different protocols at the link and/or application layers;
- may not be aware of the true nature of the nodes providing services and resources;
- may be geographically distant.

4.3.1.2 Service and Resource Discovery

Service discovery aims at obtaining the hostport of the CoAP servers in the network, while resource discovery is discovery of the resources that a CoAP server manages. Because of its role in managing the life-cycle of nodes residing in its network, the IoT gateway is naturally aware of the presence of the nodes and the available services and resources. When the IoT gateway detects that a CoAP node has joined its IP network, it can query the CoAP node, asking for the list of provided services; in CoAP this is done by sending a GET request to the /.well-known/core URI. Such information (the resource directory) is then locally maintained by the IoT gateway and successively used to route incoming requests to the proper resource node. According to this mechanism, the IoT gateway may act as an RD for the CoAP nodes within the network.

In Section 4.3.2, we detail how IoT gateways can be federated in a P2P overlay in order to provide a distributed and global service and resource directory that can be used to discover services at a global scale. In Section 4.3.3, we then provide a zero-configuration solution for discovery of resources and services within a local scope, with no prior knowledge or intervention required on any node of the network. This allows the IoT gateways to populate and update their resource and service directories.

4.3.2 A P2P-based Large-scale Service Discovery Architecture

As stated in Section 4.3.1, IoT gateways can be federated in a P2P overlay in order to provide a large-scale service discovery mechanism. The use of a P2P overlay can provide several desirable features:

- *scalability*: P2P systems are typically designed to scale and increase their capacity as the number of participants increases;
- *high-availability*: P2P systems are inherently robust because they have no single point of failure and the failure of a node does not compromise the overall availability of the services and resources provided;
- *self-configuration*: P2P systems provide mechanisms to let the overlay re-organize itself automatically when nodes join and leave, requiring no direct intervention for configuration.

These features fit perfectly in IoT scenarios, where billions of objects are expected to be deployed. Among several approaches to implementing P2P overlays, structured overlays, such as DHTs, have some interesting features, including efficient storage and lookup procedures, resulting in deterministic behavior. On the contrary, with unstructured overlays, flooding techniques are used for message routing. In the remainder of this section, we propose a P2P-based approach that provides a scalable and self-configuring architecture for service discovery at a global scale.

IoT gateways are organized as peers of a structured P2P overlay, which provides for efficient name resolution for CoAP services. The large-scale service discovery architecture presented in this work relies on two P2P overlays:

- the Distributed Location Service (DLS) [84]
- the Distributed Geographic Table (DGT) [85, 86].

The DLS provides a name resolution service to retrieve all the information needed to access a resource (of any kind) identified by a URI. The DGT builds up a distributed geographical knowledge, based on the location of nodes, which can be used to retrieve a list of resources matching geographic criteria. The combination of these two P2P overlay systems allows for the building of a distributed architecture for large-scale service discovery, with the typical features of P2P networks (scalability, robustness, and self-configuration), yet enabling the unique feature of service and resource discovery on a geographical basis. In the following, we first detail the DLS and DGT and then we describe the overall envisioned system architecture.

4.3.2.1 Distributed Location Service

The DLS is a DHT-based architecture that provides a name resolution service based on storage and retrieval of bindings between a URI, identifying resources (e.g., web services) and the information that indicates how they can be accessed [84]. In essence, the DLS implements a location service that can be used to store and retrieve information for accessing services and resources. Together with each contact URI, other information can be stored, such as the expiration time, an access priority value, and, optionally, a human-readable text (e.g., a contact description or a name).

The service provided by DLS can be considered as similar to that of the DNS, since it can be used to resolve a name to retrieve the information needed to access the content related to that name. However, the DNS has many limitations that the DLS overcomes, such as:

- the DNS applies only to the fully qualified domain names (FQDN) and not to the entire URI;

- the DNS typically has long propagation times (further increased by the use of caching), which are not suited to highly dynamic scenarios, such as those encompassing node mobility;
- the DNS essentially provides the resolution of a name, which results in an IP address, but it does not allow for storage and retrieval of additional useful information related to the resolved URI, such as the description and the parameters of the hosted service.

Another important feature that makes the use of the DLS preferable to the DNS is its robustness. If a DNS server is unreachable, then resolution cannot be performed. In contrast, P2P overlays do not have single point of failure that might cause service disruption, resulting in a more robust, dynamic, and scalable solution.

A DLS can be logically accessed through two simple methods:

- *put(key,value)*;
- *get(key)*;

where *key* is a resource URI (actually its hash), while *value* is structured information that may include location information (e.g. a contact URI) together with a display name, expiration time, priority value, etc. The *get(key)* method should return the set of the corresponding values (actually the contact information) associated with the targeted resource. The removal of a resource is performed by updating an existing resource through a *put* operation with expiration time set to zero. This mapping allows the approach to support:

- *mobility*: it is sufficient to put and replace an old resource with an updated one that considers the new position of the resource;
- *replication*: it is sufficient to execute several *put* operations for the same resource in order to have multiple replicas diffused in the DHT.

The DLS interface can be easily integrated with existing networked applications, such as a middleware layer offering services to applications and working on top of standard transport protocols. Different RPC protocols, such as dSIP [87] and RELOAD [88], may be used for messaging, regardless of the actual selected DHT algorithm (e.g., Chord or Kademlia).

4.3.2.2 Distributed Geographic Table

The DGT [85, 86] is a structured overlay scheme, built directly using the geographical location of the nodes. Unlike DHTs, with a DGT each participant can efficiently retrieve node or resource information (data or services) located near any chosen geographic position. In such a system, the responsibility for maintaining information about the position of active peers is distributed among nodes, so that a change in the set of participants causes a minimal amount of disruption.

The DGT is different from other P2P-based localization systems, where geographic information is routed, stored, and retrieved among nodes organized according to a structured overlay scheme. The DGT principle is to build the overlay by directly taking into account the geographic positions of nodes. This allows for building of a network in which overlay neighbors are also geographic neighbors; no additional messages are needed to identify the closest neighbors of a peer. The main difference between the DGT and the DHT-based P2P overlays is the fact that the DGT overlay is structured in such a way that the messages are routed exclusively according to the geographic locations of the nodes, rather than on keys that have been assigned to the nodes. Typically, DHTs arrange hosts at unpredictable and unrelated points in the overlay, deriving keys through hashing functions. In contrast, the DGT ensures that hosts that are geographically close are also neighbors in the overlay.

The DGT provides a primitive *get(lat, lon, rad)*, which returns a list of nodes that fall inside the circular region centered at *(lat, lon)* with radius *rad*. Each node that provides a service can be looked up. The *get* primitive is used to localize the list of nodes in a certain geographic region. It might be possible to extend the *get* primitive by introducing query filters, making it possible to return only matching services. The DGT does not provide a generic *put* primitive that can be invoked on the overlay as a whole. However, it is possible to extend the classical DGT behavior with a generic *put* primitive, consisting of the detection of a list of peers in a given area (through the native DGT *get* primitive) and, subsequently, to invoke a *put* method directly on each of the detected peers.

4.3.2.3 An Architecture for Large-scale Service Discovery based on Peer-to-peer Technologies

The mechanisms presented in the previous subsections are the key ingredients of a large-scale service discovery architecture. In Figure 4.4, an illustrative representation of the system architecture is shown. Several IoT gateways managing their respective networks are interconnected through the two P2P overlays. Each IoT gateway is,

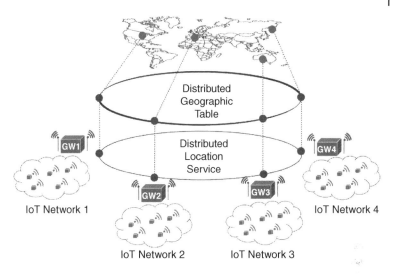

Figure 4.4 Large-scale service-discovery architecture. IoT gateway nodes act as peers of two different P2P overlays. The DLS overlay is used for discovering resources and services: a "white-pages" service that provides a name resolution service to be used to retrieve the information needed to access a resource. The DGT is used as a "yellow-pages" service, for learning about the existence of IoT gateway nodes in a certain geographical neighborhood.

at the same time, a DLS peer and a DGT peer. The data structures of the overlays are separated, since they pertain to different operations of the overall architecture. The DLS and DGT overlays are loosely coupled. The IoT gateway uses the DLS to publish/lookup the details of resources and services, and the DGT to publish its presence or discover existing IoT gateways in a given geographic area. This separation allows the IoT gateway to access the services provided by each overlay as a "black-box", without any risk of direct interference between the overlays. The IoT gateway is responsible for implementing the behavior required by the service discovery architecture.

The lifecycle of an IoT gateway is shown in Figure 4.5 and can be described as follows:

- Upon start up, the IoT gateway joins the DLS and DGT overlays.
- The IoT gateway publishes its presence in the DGT by issuing a DGT.put (lat, lon, URI_{GW}) request.
- When the IoT gateway detects a new CoAP node in the network, through any suitable means (e.g., Zeroconf), it fetches the node's

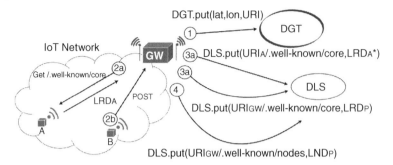

Figure 4.5 Messages exchanged when a new node joins the network. First, the IoT gateway discovers the resources of a new CoAP server or stores them on behalf of a CoAP client. Finally, DGT and DLS are updated with information about the new node.

local resource directory (LRD) through a CoAP GET request targeting the /.well-known/core URI. The LRD is filled with documents in JSON-WSP[3] or similar formats (such as CoRE Link Format) containing the description of all the resources that are hosted by the CoAP node and the information to be used to access them. At this point, the resources included in the fetched node's LRD are added to the IoT gateway's LRD.

- If the IoT gateway is willing to let the resources be reachable through it, it will modify its LRD to include the references of the URLs to be used to reach the resources through the IoT gateway, obtaining a new LRD, denoted as LRD*; the IoT gateway could also delete from the LRD all the references directly related to this resource, in order to avoid having a resource that could be accessed without the IoT gateway relaying messages.
- The IoT gateway publishes the LRD* in the DLS through a *DLS.put(URI$_{node}$/.well-known/core,LRD*)* request.
- The IoT gateway keeps track of the list of nodes that are in its managed network, by adding the node to a local node directory (LND).
- The IoT gateway publishes the LND pair in the DLS through a *DLS.put(URI$_{GW}$/.well-known/nodes,LND)* request.

---

3 JavaScript Object Notation Web-Service Protocol (JSON-WSP) is a web-service protocol that uses JSON for service description, requests, and responses. It has been designed to cope with the lack of service description specification with documentation in JSON-RPC, a remote procedure call protocol in JSON format.

- If, in addition, the IoT gateway acts as origin server, it stores its own resources, which will then be published as soon as it receives CoAP POST requests from CoAP clients residing in the inner network.

Steps 3 to 7 are repeated for each CoAP node detected in the network. By publishing all the LRDs in the DLS, a distributed resource directory (DRD) is obtained. The DRD provides global knowledge of all the available resources. The use of LNDs provides a census of all the nodes that are within a certain network. Location information is managed with JSON-WSP or CoRE Link Format documents, which provide all the details related to parameters and return values. This is similar to WSDL documents, but in a more compact, yet just as descriptive, format than XML. As soon as a node joins a local network and discovers the presence of an IoT gateway (it can be assumed that either the IoT gateway address is hard-coded or the node joins the RPL tree, finding the IoT gateway – other mechanisms may also be possible), the node announces its presence. We note that this phase is optional, in the sense that other discovery mechanisms can be adopted. When the IoT gateway detects this advertisement, it issues a GET /.well-known/core to the node, in order to discover its available resources. The node, in return, replies by sending a JSON-WSP or CoRE Link Format document describing its exposed resources, the URI to access them, and their data format. Finally, the IoT gateway will parse this response and will populate the DLS and DGT accordingly. If other IoT gateways are present within a certain network, they can act as additional access points for a resource: this can be achieved by publishing a LRD*' containing the URLs related to them. This will lead to highly available and robust routing in very dynamic scenarios where IoT gateways join and leave the network. Should one want to provide fault-tolerance, information replication mechanisms can be also introduced [89].

In the proposed architecture, the DLS can be interpreted as a "white-pages" service to resolve the name of a service, in the form of a URI, to get all the information needed to access it. Similarly, the DGT can be interpreted as a "yellow-pages" service, used to retrieve a list of available services matching geographic location criteria; that is, in the proximity of a geographic position. Note that the DGT is just one possible solution to get matching services; other mechanisms might be adopted. These might not be related to geographic locations, but instead on different matching criteria, such as taxonomies/semantics.

This is the case if the search is by type of service rather than by geographical location.

The distinction between the lookup services provided by DLS and DGT avoids the inclusion, in the URI, of service or resource information that can dynamically change (such as the location), thus making it possible to support mobility of services and resources. The DGT and the DLS run in parallel, and the IoT gateways of a IoT sub-network act as peers of both the DLS and the DGT. The resulting architecture is very flexible and scalable: nodes that may join and leave the network at any time. In fact, as explained in the previous subsections, the nature of DLS and DGT P2P overlay networks allows new IoT gateways to be added without requiring the re-computation of the entire hash table. Vice versa, only the nodes responsible for maintaining the resources close to the joining node must update their hash tables in order to include the resources of the new node.

A client needing to retrieve data from a resource and with no information about the URI to contact, must perform the operations shown in Figure 4.6. It can perform service discovery through the mediation of a known IoT gateway that is part of the DLS and DGT overlays. The procedure can be detailed as follows (the first five steps are explicitly shown in Figure 4.6):

1) The client contacts a known IoT gateway in order to access the DLS and DGT overlays for service discovery.

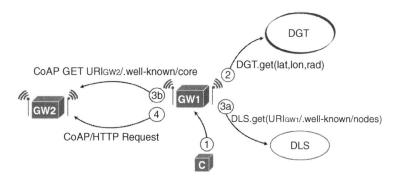

Figure 4.6 Data retrieval operations: 1) the client C contacts a known IoT gateway GW1; 2) GW1 accesses the DGT to retrieve the list of IoT gateways available in a given area; 3a) GW1 selects one of these IoT gateways, namely GW2; 3b) GW1 discovers the nodes managed by the GW2 through the DLS or directly by contacting GW2; 4) finally, GW1 queries the node, associated with the resource of interest, managed by GW2.

2) The client uses the DGT to retrieve a list of IoT gateways that are in the surroundings of a certain geographical location through a *DGT.get(lat, lon, rad)* request.

3a) The IoT gateway selects one of the IoT gateways returned by the DGT and discovers the list of its managed nodes, through a *DLS.get(URI$_{GW1}$/.well-known/nodes)* request.

3b) The IoT gateway discovers the resources that are reachable:
   - by executing a *DLS.get(URI$_{node}$/.well-known/core)* procedure or
   - by issuing a CoAP GET request for *URI$_{GW2}$/.well-known/core.*

4) The IoT gateway interacts with the resource by issuing CoAP or HTTP requests targeting the selected resource through the appropriate IoT gateway. The client can then contact the URI of the resource, either directly through CoAP (if supported by the IoT gateway) or by HTTP (by delegating to the IoT gateway the HTTP-CoAP request translation).

5) Once the command has been transmitted to the CoAP server, the latter will reply with the requested data.

6) If supported, the response will be through CoAP to the client. Otherwise, the IoT gateway will be in charge of response translation.

4.3.3 Zeroconf-based Local Service Discovery for Constrained Environments

Service discovery within a local network can be performed using several mechanisms. In scenarios where a huge number of devices are involved or external human intervention is complicated, it is desirable that all devices can automatically adapt to the surrounding environment. The same considerations apply to devices that do not reside in a particular environment but are characterized by mobility, for example smartphones. In both cases, a service discovery mechanism, which requires no prior knowledge of the environment, is preferable. In this section, we propose a novel lightweight Zeroconf-based mechanism for service and resource discovery within local networks.

4.3.3.1 Architecture
Our local service discovery mechanism is based on the Zeroconf protocol suite. It involves the following elements:

- IoT nodes (smart objects) belonging to an IoT network;
- an IoT gateway, which manages the IoT network and acts as the RD;
- client nodes, which are interested in consuming the services offered by the IoT nodes.

We assume that IP multicast is supported within the local network and that DHCP [90] provides the dynamic configuration for the IP layer.

4.3.3.2   Service Discovery Protocol

There are essentially two relevant scenarios for the application of the proposed service discovery protocol:

- a new device offering some service is added to the network and starts participating actively;
- a client, which is interested in consuming the services offered by the nodes already present in the network, discovers the available services.

In the former scenario, the procedure for adding a new service to the network can be performed in two different ways, depending whether:

- the smart object can be queried for its services (using the /.well-known/core URI); or
- it posts the information related to the services it is offering on the IoT gateway, which acts as a resource directory.

The difference between the two scenarios also involves the characterization of the smart object as a CoAP server or as a CoAP client, respectively. If the device acts as a CoAP (origin) server, the service discovery procedure, as shown in Figure 4.7, is the following:

1) The IoT node joins the network and announces its presence by disseminating a mDNS message for a new service type *_coap._ udp.local.*
2) The IoT gateway, listening for events related to service type *_coap._udp.local.*, detects that a new node has joined the network.
3) The IoT gateway queries the new node for its provided services by sending a CoAP GET request targeting the URI /.well-known/core.
4) The IoT node replies with a JSON-WSP or CoRE Link Format document describing the offered services.
5) The IoT gateway updates the list of services that it manages on behalf of the constrained nodes residing in the network, thus making these services consumable by clients residing outside of the IoT network (e.g., remote Internet hosts, which may be unaware of the constrained nature of the network where the service of interest is located).

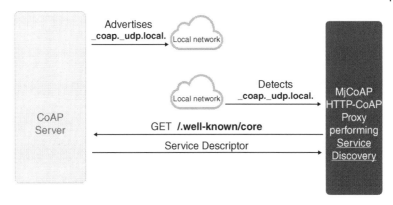

Figure 4.7 Service advertisement by CoAP server detected by HTTP-to-CoAP proxy.

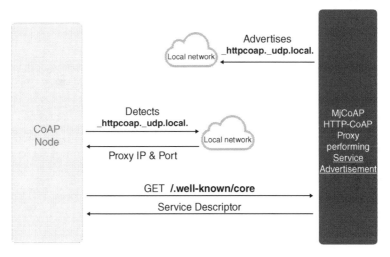

Figure 4.8 Service advertisement by HTTP-to-CoAP proxy detected by CoAP client.

If the device acts as a CoAP client, on the other hand, the service discovery procedure, as shown in Figure 4.8, is the following:

1) The proxy, which is a module of the IoT gateway, announces its presence periodically, by disseminating a mDNS message for a new service type _httpcoap._udp.local_.

2) The joining smart object, which is listening for events related to the service type advertised by the IoT gateway (*_httpcoap._udp.local.*), detects that an IoT gateway is available in the network.

3) The smart object sends a CoAP GET request to the URI /.well-known/core to get a description of the services that the IoT gateway provides and other information that might be used to detect the most suitable proxy for the client.

4) The IoT gateway replies with a JSON-WSP or CoRE Link Format document describing the services it provides.

5) The smart object processes the payload and then sends a CoAP POST/PUT request to the IoT gateway to store resources to be made available to external clients.

In this scenario, the IoT gateway does not simply forward incoming requests and relay responses, but it acts as a server both towards

- the generator of the resource (CoAP client) from which it receives CoAP POST requests;
- external clients, to which it appears as the legitimate origin server, since the generator of the data is not a CoAP server.

When a client needs to discover the available services, the procedure comprises the following steps:

1) The client sends a CoAP or HTTP request to the proxy targeting the URI /.well-known/core.

2) The proxy replies with a JSON-WSP or CoRE Link Format document describing all the services managed on behalf of the nodes;

3) The client then uses the received information to perform subsequent CoAP or HTTP requests in order to consume the required services.

The use of IP multicast (i.e., mDNS) has the chief advantage of avoiding having to set a priori the actual network address of any device present, thus eliminating the need for any configuration.

4.3.4 Implementation Results

The solutions presented Sections 4.3.1–4.3.3 may be used for many large IoT scenarios in which scalable and reliable service and resource discovery is required. In particular, we focus on a smart-infrastructure surveillance scenario, where given areas of interest can be monitored by means of wireless devices. Each device (smart object) is characterized by the type of the collected data and by its position.

A system user may then be interested either in directly contacting a given resource (e.g., a sensor) or having the list of all available resources in a given area. Such wireless sensors are grouped in low-power wireless networks with one or more gateways acting as interfaces between the resulting constrained wireless network and the rest of the network (namely, in the considered scenario, the Internet).

In order to validate the feasibility of the proposed solution and to evaluate its performance, extensive experimentation has been carried out in the reference smart-infrastructure surveillance scenario. The performance evaluation focuses on both local and large-scale service discovery mechanisms, as described in Sections 4.3.2 and 4.3.3, respectively.

4.3.4.1 Local Service Discovery

The first phase of the experimental performance analysis focuses on the discovery of new CoAP services (associated with constrained devices) available in the local network.

The performance evaluation of our Zeroconf-based local service discovery strategy was conducted using Zolertia Z1 Contiki nodes, simulated in the Cooja simulator. The Contiki software stack running on each node was configured so as to fit in the Z1's limited available memory, in terms of both RAM and ROM – Z1 nodes feature a nominal 92 kB ROM (when compiling with 20-bit architecture support) and an 8 kB RAM. In practice, the compilation with the Z1 nodes was performed with a 16-bit target architecture, which lowers the amount of available ROM to roughly 52 kB. The simulated smart objects run Contiki OS, uIPv6, RPL, NullMAC, and NullRDC. The software stack deployed on the smart objects includes our lightweight implementation of the mDNS [82] and DNS-SD [83] protocols, developed in order to minimize memory footprint and to include all the needed modules in the smart objects. The implementations comply with the IETF standards defined in the RFCs and can be replaced by any other compatible implementation, should no particular constraint on the code size be present. The local service discovery mechanism was tested on IEEE 802.15.4 networks formed by Contiki nodes arranged in linear and grid topologies. The performance indicators were:

- the time needed to perform a DNS-SD query – from the DNS-SD client perspective;
- the time needed to process an incoming DNS-SD query and respond – from the DNS-SD server perspective.

Table 4.1 Local service discovery metrics.

Metric	Description	Dimension
QC	Query client time: the time needed by a node acting as client to send a DNS-SD query and receive a response	ms
QS	Query server time: the time needed by a node acting as server to construct and send a response back to a DNS-SD client	ms

The impact of the number of constrained nodes (and, therefore, the number of hops needed) in the network was analyzed. All the results were obtained by performing 100 service discovery runs on each configuration. The specific performance metrics are detailed in Table 4.1.

In Figure 4.9a, the considered linear topology, with a maximum of 20 nodes deployed in Cooja, is shown. In particular:

- node 1 is the 6LoWPAN border router (6LBR), which is the root of the RPL tree;
- node 2 is the node acting as DNS-SD server;
- node 3 is the node acting as DNS-SD client.

The distance between nodes was set so that the query had to follow a multi-hop path consisting of as many hops as the number of nodes in the network. In Figure 4.9b, the corresponding performance, in terms of QC/QS times, as functions of the number of smart objects, is shown. The QS time has a nearly constant value of around 65 ms, since the processing time is independent of the number of nodes in the network. The QC time is a linear function of the number of hops (which, in our scenario, coincides with the number of nodes), since the query packet has to be relayed by each intermediate node to reach the DNS-SD server node.

More complex bi-dimensional topologies were also tested in order to evaluate grid-like deployments. Different sizes and arrangements for grids were considered, as shown in Figure 4.10. In all cases:

- node 1 is the 6LBR;
- node 2 is the node acting as DNS-SD server;
- node 3 is the node acting as DNS-SD client.

The topologies in Figure 4.10 are: (a) Grid-A (3 hops); (b) Grid-B (4 hops); (c) Grid-C (6 hops); (d) Grid-D (5 hops). The corresponding

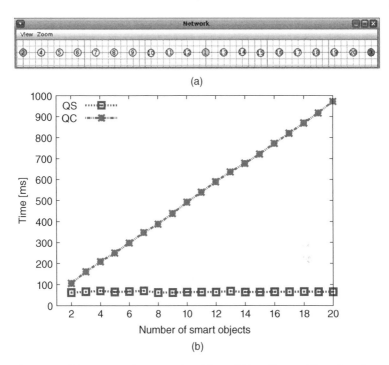

Figure 4.9 (a) Linear topology considered for multi-hop Zeroconf-based service discovery; (b) average time of Zeroconf-based service discovery on Contiki nodes with linear topology.

performance of service resolution, in terms of QC/QS times, is shown in Figure 4.11. Just like in the linear case, the QS time is independent of the network size (around 65 ms were still needed by the DNS-SD server-side processing). As the number of nodes participating in the network increases, the QC time increases as well, because of the need for multi-hop communications from client to server. It can be seen that, in the case of Grid-D, even though the number of nodes is larger than in the case of Grid-C, the QC time is shorter. This is because the distance between the nodes has decreased from 40 m to 30 m (to minimize collisions due to the use of NullMAC) and, therefore, the total number of hops from the client to the server decreases. In general, it can be concluded that, at a fixed node density, the QC time is a linear function of the number of hops.

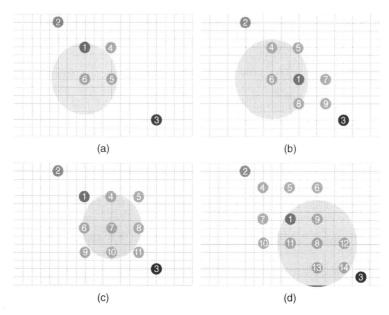

(a)   (b)

(c)   (d)

Figure 4.10 Grid topologies considered for bi-dimensional deployments of smart objects: (a) Grid-A (3 hops); (b) Grid-B (4 hops); (c) Grid-C (6 hops); (d) Grid-D (5 hops).

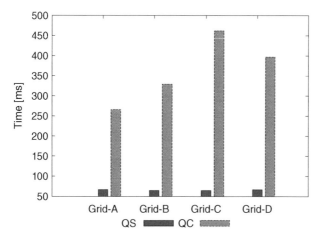

Figure 4.11 Average QC/QS times of the Zeroconf-based service discovery in the grid topologies shown in Figure 4.10.

4.3.4.2 Large-scale Service Discovery

The second performance evaluation phase focuses on a P2P overlay in which multiple IoT gateways join the network in order to store new resouces in the DLS overlay and retrieve references to existing ones. The aim of this evaluation was to test the validity of the proposed approach with different configurations and, in particular, to measure the average time required by an IoT gateway to complete the three main actions in the network (JOIN, PUT and GET) for different sizes of the P2P overlay. We focus only on the evaluation of the DLS overlay since the published content pertains to IoT services and resources and, therefore, it represents the component of the proposed service discovery architecture that is directly related to IoT services and resources. The DGT allows for a structured geographical network that can be used to efficiently discover available nodes based on location criteria in a content-agnostic way; this is what the DGT was designed and thoroughly evaluated for, both in simulative environments and real-world deployments [91, 92].

The performance evaluation was carried out for several configurations, with different numbers of IoT gateways (which are also the peers of the overlay). Each IoT gateway acts as boundary node of a wireless network with CoAP-aware sensor nodes. The DLS overlay uses a Kademlia DHT and the dSIP protocol for P2P signalling [87, 93], both implemented in Java. The P2P overlay contains up to 1000 nodes deployed over an evaluation platform comprising four cluster hosts, each an 8-CPU Intel®Xeon®E5504 running at 2.00 GHz, with 16~GB RAM and running the Ubuntu 12.04 operating system. The number of nodes in the P2P network was split evenly among all cluster hosts (up to 250 peers per cluster host), which were connected using a traditional switched Ethernet LAN. The HTTP-to-CoAP proxy functionality relies on two different implementations:

- one based on the mjCoAP library [23], an open-source Java-based RFC-compliant implementation of the CoAP protocol;
- the other based on the Californium platform [94].

Both HTTP-to-CoAP proxies were written in Java and provide their own local service discovery mechanisms. The use of two different types of HTTP-to-CoAP proxy shows clearly how the overlay can be easily developed and integrated with currently available technologies. The sensor nodes are either Arduino boards or Java-based emulated CoAP nodes (just for emulating large network scenarios).

Each performance result is obtained by averaging over 40 executions of PUT and GET procedures for each size of the overlay.

As anticipated, the following performance metrics are of interest:

- elapsed time for a JOIN operation (dimension: [ms]);
- number of rounds for PUT operations (adimensional);
- number of rounds for GET operations (adimensional).

The selection of the number of rounds for PUT and GET operations, rather than their times, is expedient to present performance results that are independent of the actual deployment environment. For the JOIN operation, the average total time required to completion is shown in order to provide a practical measurement of the complexity of this operation. However, the very nature of all operations relies on a common iterative procedure [78], thus making it possible to intuitively derive the behavior of all operations in terms of time and rounds.

The performance results are shown in Figure 4.12. As expected, the complexity, in terms of JOIN time and numbers of rounds for PUT/GET operations, is a logarithmically increasing function of the number of peers. In Figure 4.12, the experimental data are directly compared with the following logarithmic fitting curves [95]:

$$\text{Join time} \simeq 16.5 + 61.29 \cdot \log n$$

$$\text{\# of rounds PUT} \simeq -5.75 + 3.44 \cdot \log n$$

$$\text{\# of rounds GET} \simeq -0.40 + 0.15 \cdot \log n.$$

This clearly proves the scalability brought by the use of a P2P approach, confirming the formal analysis and results of Maymounkov and Mazières [78].

To summarize, we have presented a novel architecture for self-configurable, scalable, and reliable large-scale service discovery. The proposed approach provides efficient mechanisms for both local and global service discovery. First, we have described the IoT gateway and the functionalities that this element must implement to perform resource and service discovery. Then, we have focused on large-scale distributed resource discovery, exploiting a proper P2P overlays, namely DLS and DGT, which implement, respectively, "white-pages" and "yellow-pages" services. Finally, we have shown a solution for automated local service discovery that allows for discovery of resources available in constrained WSNs and their publication into the P2P overlay with no need for any prior configuration (Zeroconf).

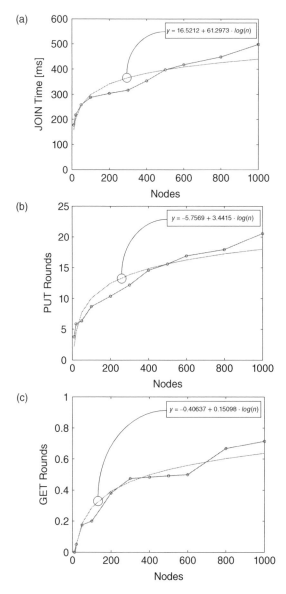

Figure 4.12 Experimental results collected to evaluate the performance of the DLS overlay: (a) average elapsed time for JOIN operations, (b) the average number of rounds (adimensional) for PUT operations, and (c) the average number of rounds (adimensional) for GET operations on the DLS towards the number of active IoT gateways in the P2P network. Plotted data have also been used to construct fitted curves (in red); the formulae of which are reported in the top-right hand corners.

Extensive experimental performance evaluation of the proposed local and large-scale service discovery mechanisms was performed. For the local service discovery mechanism, experiments were conducted on Contiki-based nodes operating in constrained (IEEE 802.15.4) networks with RPL in the Cooja simulator. The large-scale service discovery mechanism was deployed and tested on P2P overlays of different sizes, spanning from a few to 1000 peers, in order to evaluate the performance in terms of scalability and self-configuration. The results show that the time required for service resolution in the Zeroconf-based approach for local service discovery is linearly dependent on the number of hops in the path between the client and server node. For large-scale service discovery, the adoption of a P2P overlay provides scalability in terms of the time required to perform the basic publish/lookup operations.

In conclusion, the easy and transparent integration of two different types of overlays shows the feasibility and reliability of a large-scale architecture for efficient and self-configurable service and resource discovery in IoT networks.

4.4 Lightweight Service Discovery in Low-power IoT Networks

Zeroconf [60] is a protocol suite which reuses the semantics of DNS messages over IP multicast to provide name resolution and service discovery/advertisement over local networks. In order to support Zeroconf service discovery mechanisms, it is very important that the network supports IP multicasting and implements proper forwarding techniques to guarantee that packets are delivered to all group nodes and avoids the establishment of loops. Using efficient packet forwarding mechanisms can bring benefits in multi-hop communications among smart objects, in terms of delay and energy consumption. Moreover, it is also important to note that the limited amount of memory available on smart objects requires the adoption of small-footprint mechanisms, in order to allow developers to integrate a complete software stack, without having to sacrifice some modules in order to meet the memory constraints. Although the IETF ROLL working group is defining a Multicast Protocol for Low power and Lossy Networks (MPL) [96], based on the Trickle algorithm [97], some applications might have different requirements and could benefit from the adoption of other multicast techniques.

In the following sections we present a lightweight and low-power multicast forwarding protocol for service discovery in smart objects operating in IEEE 802.15.4 multi-hop networks. The proposed solution features a smaller memory footprint than in other state-of-the-art solutions. The proposed mechanism has been implemented on Contiki OS-enabled smart objects. Extensive testing is carried out in the Cooja simulator to evaluate the feasibility and efficiency, in terms of delay and energy consumption, of the proposed mechanism.

Local service discovery mechanisms in LANs have been proposed in the literature. Protocols like UPnP [57] and SLP [58, 59] focus on automatic announcement and discovery of in-network existing services. However, their porting to IoT devices is not straightforward because of the severe computation and energy constraints of the nodes. An alternative to these protocols relies on multicast forwarding. For instance, Jung and Kastner proposed an efficient group communication strategy for the CoAP and the Efficient XML Interchange protocols [98]. To achieve group communication, they rely on the Open Building Information eXchange standard. However, this implementation runs on Raspberry PI nodes, so it is not suitable for constrained devices.

Concerning 6LoWPAN and IPv6, the only active IETF draft on efficient multicast forwarding is MPL [96], that relies on the Trickle algorithm to manage transmissions for both control and data plane. The different multicast interfaces, identified by an unicast address and associated with one or more multicast domains, are handled separately, so as to maintain an independent seed set to decide whether to accept a packet or not. The MPL forwarder, which is in charge of sending data messages, has two different possible strategies: proactive or reactive. In the former case, the MPL forwarder schedules the transmission of MPL data messages using the Trickle algorithm, without any prior indication that neighbor nodes are yet to receive the message. After transmitting a limited number of MPL data messages, the MPL forwarder may terminate proactive forwarding for the MPL data message. In the latter, the MPL forwarder sends link-local multicast MPL control messages using the Trickle algorithm. MPL forwarders use MPL control messages to discover new MPL data messages that have not yet been received. When an MPL forwarder discovers that a neighbor MPL forwarder has not yet received an

MPL data message, it schedules the transmission of those MPL data messages using the Trickle algorithm. The two approaches can coexist at the same time.

Oikonomou and Phillips proposed Stateless Multicast RPL Forwarding (SMRF [99]), which relies on the presence of the RPL routing protocol and requires group management information to be carried inside RPL destination advertisement object (DAO) messages. However, since, for our goal, a less complicated multicast strategy (no group management is required) is needed, we prefer to rely on a more lightweight flooding technique, which adapts well to duty-cycled devices operating in RPL networks implementing the Zeroconf protocol suite.

4.4.1 Efficient Forwarding Protocol for Service Discovery

Zeroconf is a protocol that allows for automatic creation of computer networks based on the TCP/IP Internet stack. It does not require any external configuration [60]. Zeroconf provides three main functionalities:

- automatic network address assignment;
- automatic distribution and resolution of host names;
- automatic location of network services.

Automatic network assignment comes into the picture when a node first connects to the network. The host name distribution and resolution is implemented using multicast DNS (mDNS) [61], a service that has the same interfaces, packet formats, and semantics as standard DNS, so as to resolve host names in networks that do not include a local name server. Zeroconf also allows to for publication of services (DNS-SD) in a local network. Both mDNS and DNS-SD do not require the presence of any server (and, therefore, its knowledge) to perform publish, lookup, and name resolution operations, but rely on the use of IP multicast communications in order to address all the nodes in the local network. Zeroconf specifies that mDNS and DNS-SD messages (for both requests and responses) must be sent to the mDNS IPv4/IPv6 link-local multicast address (i.e., 224.0.0.251 and ff02::fb, respectively). However, Zeroconf does not require per-group multicast routing: according to the protocol specifications, messages should simply reach all nodes in the local network.

6LoWPAN defines methods

- to transmit IPv6 packets and
- to form IPv6 link-local addresses and statelessly autoconfigure addresses on IEEE 802.15.4 networks.

The RPL protocol defines a routing protocol for IP communications in LLNs. The IETF ROLL Working Group is working on the definition of MPL, a multicast protocol providing IPv6 multicast forwarding in constrained networks. This could become a general multicast technique able to manage multicast groups of any size. However, in some scenarios, such as Zeroconf service discovery, there is no need to actually adopt such a full-feature multicast protocol. For the sake of Zeroconf service discovery, it is sufficient to provide a multicast forwarding mechanism that guarantees that messages can be delivered to all nodes in the local network. In this section, we detail a simple and efficient forwarding algorithm that can be adopted by constrained devices operating in RPL networks with ContikiMAC radio duty-cycling protocol, in order to enable IP multicast communications with a small footprint, targeting Zeroconf service discovery.

4.4.1.1 Multicast through Local Filtered Flooding

Flooding is the simplest routing protocol for broadcasting a packet to all nodes in the network. From a practical implementation point of view, each node forwards a received packet to all its neighbors. This technique is effective only for cycle-free topologies (i.e., trees). In the presence of graphs with cycles, it is necessary to implement duplicate detection techniques to avoid forward loops. An illustration is shown in Figure 4.13.

In order to implement an efficient mechanism to detect already-processed packets (and, thus, avoid redundant forwarding), we propose the adoption of Bloom filters [100]. Bloom filters are probabilistic data structures that can be used to add elements to a set and to efficiently check whether an element belongs to the set or not. Bloom filters provide two primitives:

- *add(x)*: add element x to the set;
- *query(x)*: test to check whether element x is in the set.

The filter does not provide a *remove(x)* primitive, so it is not possible to undo an insertion. Bloom filters are slower when performing *check*

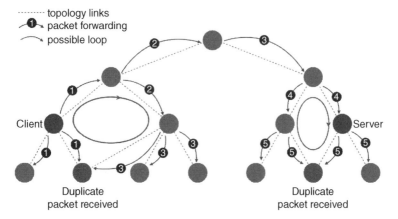

Figure 4.13 Flooding of a DNS-SD query in generic topology with cycles.

operations than equivalent probabilistic data structures (in terms of provided functionalities), such as quotient filters [101], but occupy less memory. As available memory on smart objects is extremely limited, one of the design goals of the proposed forwarding algorithm is to keep the memory footprint (both in terms of RAM and ROM) as low as possible. Therefore, Bloom filters have been selected as the most appropriate data structure to keep track of already-processed packets.

A Bloom filter is initially an array of m bits, all set to zero. The *add(x)* operation passes the input argument x through k different hashing functions and obtains k indexes in the bit array of the Bloom filter that will be set to one. The *query(x)* operation verifies whether the indexes corresponding to x are all set to one. The Bloom filter is probabilistic in the sense that a *query(x)* operation can return false positives: there can exist two values x_1 and x_2, such that $query(x_1) = query(x_2) = true$. False negatives, on the other hand, are not possible: this means that if a *query(x)* returns *false*, then x is not in filter. The *query(x)* operation can thus return either "probably in the set" or "not in the set".

Bloom filters can be instantiated to meet specific application requirements by selecting the parameters m (number of bits in the array) and k (number of hashing functions). For instance, the choice of m and k has an impact on the probability of getting false positives for *query(x)* operations and on memory occupation. In any case, the impossibility of removing an element from the filter leads to an increase in the probability of false positives as more and more elements are added to the filter. Since the purpose of using a Bloom filter in the forwarding algorithm is to detect duplicate elements, in order to cope with the problem of false positives, the Bloom filter is periodically reset. Resetting the filter might introduce some unnecessary retransmissions if the filter is emptied before receiving a duplicate packet. However, retransmissions are preferable to packet drops in order to guarantee that a multicast packet reaches all hosts. Moreover, such unnecessary retransmissions might occur no more than once, as the packet would then be added to the filter and not processed upon future receptions. To summarize, upon receiving a packet, a node will perform the following steps:

1) Check if the incoming IP packet has already been processed, by performing a *query* operation in the Bloom filter.
2) If the Bloom filter contains the packet, discard it; otherwise, the packet is added to the Bloom filter through an *add* operation.
3) If needed, forward the received IP packet to all neighbors by means of local IEEE 802.15.4 broadcast.

4.4.2 Efficient Multiple Unicast Forwarding

While the described algorithm implements an optimized flooding mechanism by avoiding loops through the introduction of Bloom filters, broadcasting with the ContikiMAC radio duty-cycling protocol results in inefficient transmissions, leading to higher energy consumption and end-to-end delays. In fact, in ContikiMAC, a broadcasting node must repeatedly transmit a packet for the full wake-up interval [49], in order to ensure that it can be received by all neighbor nodes, regardless of their wake-up time. This conservative approach has the following drawbacks:

- the number of transmitted packets is larger than necessary, and therefore energy consumption is higher;
- when a node is broadcasting a packet, other nodes are not allowed to transmit, and this delays the transmission until the channel is clear;
- since ContikiMAC broadcasting does not make provision to acknowledge received packets, it might be that not all neighbors have successfully received the packet, thus leading to unreliable transmission.

These inefficiencies are magnified when the channel check rate (CCR) decreases, since the full wake-up interval is longer and therefore the channel is busy for longer, thus leading to even more repeated transmissions and delays. This contrasts with the assumption that lower CCR leads to lower energy consumption. In order to tackle these issues, we replace local broadcast with multiple unicast transmission. The forwarding algorithm can therefore be optimized by selecting the receiving nodes from the list of next hops, which is retrieved from the RPL routing table. In fact, ContikiMAC provides per-node-pair synchronization, which ensures that packets are sent only when the receiver is supposed to be active. The receiver is required to send an acknowledgement for the received packet, thus transmitting packets only for as long as necessary, thus leading to more reliable transmissions.

The enhanced version of the proposed multicast protocol can therefore be detailed as follows:

1) Check if the incoming IP packet has been processed already by performing a *query* operation in the Bloom filter.
2) If the Bloom filter contains the packet, discard it; otherwise, add the packet to the Bloom filter through an *add* operation.
3) Retrieve the list of next hops from the routing table.
4) If needed, forward the received IP packet to each next hop using IEEE 802.15.4 unicast communication.

An excerpt of a sequence of transmitted frames, using broadcast for a DNS-SD query, is shown in Figure 4.14.

The equivalent packet flooding with multiple unicast transmissions is shown in Figure 4.15. Transmitted packets (TX), received packets

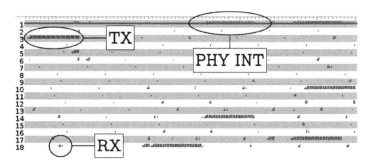

Figure 4.14 DNS-SD query propagation DNS-SD query propagation with ContikiMAC broadcast. Time is on the *x*-axis while node identifiers are on the *y*-axis.

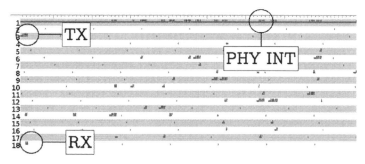

Figure 4.15 DNS-SD query propagation with multi-unicast. Time is on the *x*-axis while node identifiers are on the *y*-axis.

(RX), and PHY interference (PHY INT) are highlighted. The root of the RPL tree (node 1) is always active, while all other nodes have CCR = 8 Hz. The timelines clearly show that multiple unicast transmissions optimize the number of transmitted packets and the packet propagation delay in the network, while guaranteeing more reliable transmissions. However, this comes at the cost of a slightly increased ROM/RAM footprint.

4.5 Implementation Results

In order to evaluate the performance of the proposed multicast packet forwarding mechanism, a Contiki-based implementation has been developed. Besides the proposed multicast forwarding algorithm, the mDNS and DNS-SD protocols have been re-implemented, in order to have a smaller memory footprint than in other, already available implementations. The performance evaluation of the Zeroconf-based local service discovery strategy was conducted using WiSMote[4] Contiki nodes, simulated in the Cooja simulator. The Contiki software stack running on each node was configured in order to fit in the WiSMote's available memory, in terms of both RAM and ROM – WiSMote nodes feature a nominal 128 kB, 192 kB or 256 kB ROM and 16 kB RAM. The simulated smart objects run Contiki OS, uIPv6, RPL, and ContikiMAC.

The local service discovery mechanism was tested on Contiki nodes arranged in linear and grid topologies, as shown in Figure 4.16, in

---

4 http://wismote.org/.

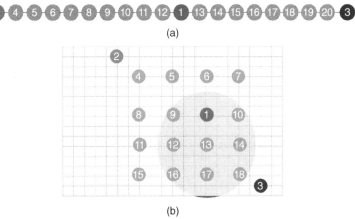

(a)

(b)

Figure 4.16 Topologies considered for Zeroconf service discovery experimentation with the proposed multicast protocol: (a) linear and (b) grid.

IEEE 802.15.4 networks, with RPL as routing protocol and Contiki-MAC as radio duty-cycling protocol. In particular:

- node 1 is the 6LoWPAN border router (6LBR), which is the root of the RPL tree;
- node 2 is the node acting as DNS-SD server;
- node 3 is the node acting as DNS-SD client.

The first performance indicator is memory occupation in terms of ROM. The proposed multicast protocol – both with broadcast and multiple unicast transmission – is compared to the MPL (with Trickle algorithm) implementation available in the Contiki 3.x fork. The results are shown in Table 4.2.

Table 4.2 ROM usage for the proposed multicast protocol and an MPL implementation: total ROM occupation and as a percentage of available memory on 128-kB WiSMote.

Library	ROM occupation [B]	Occupation of overall available ROM
This work (broadcast)	842	0.64%
This work (multiple unicast)	1454	1.11%
MPL with Trickle	3804	2.90%

As expected, the footprint of the proposed solution is significantly smaller than that of the MPL implementation: approximately 78% for broadcast and 62% for multiple unicast.

The next phase of experimentation aims at evaluating the time and the overall network energy consumption needed to perform advertisement and resolution of services using Zeroconf in the topologies shown in Figure 4.16 and the proposed broadcast-based and multiple unicast-based approaches. All the results were obtained by performing 100 service discovery runs on each configuration. The specific performance metrics are:

- query client time (QC), which is the time needed by a node acting as client to send a DNS-SD query and receive a response (dimension: [ms]);
- energy consumption (E), which is the overall network energy consumption for a DNS-SD query operation (dimension: [mJ]).

The impact of the CCR (varying from 8 Hz to 128 Hz) of the nodes participating in the constrained network is analyzed. The results for QC and E are shown in Figures 4.17 and 4.18, respectively. As expected, QC is inversely proportional to the CCR. The benefit of using multiple

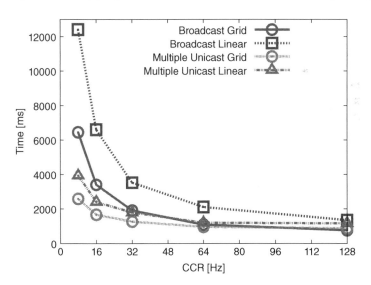

Figure 4.17 Time required to perform a DNS-SD query by a client in linear and grid topologies with broadcast and multiple unicast.

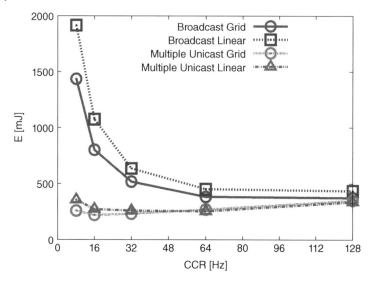

Figure 4.18 Overall network energy consumption when a client performs a DNS-SD query in linear and grid topologies with broadcast and multiple unicast.

unicast transmissions, instead of broadcast, is higher when the CCR is low, while the two approaches tend to overlap for higher values of the CCR. At typical CCR values, multiple unicast performs better than broadcast because, with lower CCR, the wake-up interval is longer and ContikiMAC broadcast transmissions occupy the channel for the whole interval. Higher CCR values mean shorter wake-up intervals and therefore other nodes in the network are likely to be blocked.

In the case of the grid topology, with a CCR of 128 Hz, broadcast is actually slightly faster than unicast. In fact, at this high rate, the smart objects are almost behaving as with null duty-cycling, which is the best-case scenario for broadcast transmission. As for energy consumption, the results clearly show that broadcast is much more inefficient than multiple unicast transmissions. It is important to point out that a significant contribution (more than 60% with multiple unicast and 30% with broadcast, at CCR = 8 Hz) to the energy consumption of the overall network comes from the border router, which does not perform duty-cycling. Again, this is motivated by the ContikiMAC broadcast strategy, which requires nodes to transmit for the whole wake-up interval. As for QC, the two approaches tend to overlap at high CCR. However, these results should not be interpreted as a suggestion to use higher CCR, as this would invalidate all the advantages of duty-cycling, which is particularly beneficial in other scenarios.

We introduced a novel multicast forwarding mechanism targeting service discovery in IoT scenarios. The proposed solution is suited to bringing efficient IP-multicast support to low-power IoT networks with duty-cycled devices. The rationale behind the presented forwarding mechanism is to have a lightweight and low-memory-footprint implementation for specific Zeroconf service discovery operations. In such scenarios, the adoption of a full-featured multicast implementation, such as MPL, might be overkill, since there is no need to provide multicast group support. Instead, an efficient flooding mechanism is used in the local network. The proposed multicast protocol relies on filtered local flooding, which adapts well to duty-cycled devices operating in LLNs with RPL. In order to avoid forward loops, we introduce Bloom filters, an efficient probabilistic data structure, to detect duplicate packets and prevent forward loops. The experimental results demonstrate that the proposed multicast protocol features a much smaller footprint, in terms of ROM occupation, than the MPL implementation available in the official Contiki fork. Finally, delay and network energy consumption have been evaluated.

5

Security and Privacy in the IoT

The Internet of Things (IoT) refers to the Internet-like structure of billions of interconnected "constrained" devices: with limited capabilities in terms of computational power and memory. These are often battery-powered, thus raising the need to adopt energy-efficient technologies. Among the most notable challenges that building interconnected smart objects brings about are standardization and interoperability. Internet Protocol (IP) is foreseen as the standard for interoperability for smart objects. As billions of smart objects are expected to appear and IPv4 addresses have mostly been used, IPv6 has been identified as a candidate for smart-object communication.

The deployment of the IoT raises many security issues, arising from

- the very nature of smart objects: the use of cryptographic algorithms that are lightweight, in terms of processing and memory requirements;
- the use of standard protocols and the need to minimize the amount of data exchanged between nodes.

This chapter provides a detailed overview of the security challenges related to the deployment of smart objects. Security protocols at network, transport, and application layers are discussed, together with the lightweight cryptographic algorithms that it is suggested should be used instead of conventional resource-hungry ones. Security aspects, such as key distribution and security bootstrapping, and application scenarios, such as secure data aggregation and service authorization, are also discussed.

Internet of Things: Architectures, Protocols and Standards, First Edition.
Simone Cirani, Gianluigi Ferrari, Marco Picone, and Luca Veltri.
© 2019 John Wiley & Sons Ltd. Published 2019 by John Wiley & Sons Ltd.

5.1 Security Issues in the IoT

Security in IoT scenarios is a crucial consideration. It applies at different levels, ranging from technological issues to more philosophical ones, such as privacy and trust, especially in scenarios like smart toys. The security challenges derive from the very nature of smart objects and the use of standard protocols. Heer *et al.* have considered the security challenges and requirements for an IP-based IoT by analyzing existing Internet protocols that might be applied to the IoT and their limitations and the problems that they might introduce [102]. Garcia-Morchon *et al.* summarize security threats in the IoT as follows [103]:

1) cloning of smart objects by unauthorized manufacturers;
2) malicious substitution of smart things during installation;
3) firmware replacement attacks;
4) extraction of security parameters (smart things may be physically unprotected);
5) eavesdropping attacks if communication channels are not adequately protected;
6) man-in-the-middle attacks during key exchange;
7) routing attacks;
8) denial-of-service attacks;
9) privacy threats.

Threats 1–4 are related to the physical nature of smart objects, which are typically deployed in public areas and cannot be constantly supervised, thus leading to potential security problems. Threats 5–8 are examples of security issues arising from the need for objects to communicate with each other. Finally, Threat 5.1 is related to the fact that smart objects might deal with personal or sensitive data, which, if intercepted by unauthorized parties, might create ethical and privacy problems.

While it is possible to cope with issues arising from the physical nature of objects only by adopting safe supply and installation measures, such as avoiding untrusted manufacturers and installers, and by trying to protect smart objects in safe places, all other security threats can be tackled by adopting means such as secure communication protocols and cryptographic algorithms. These measures enforce the following basic security properties:

- *Confidentiality*: transmitted data can be read only by the communication endpoints;
- *Availability*: the communication endpoints can always be reached and cannot be made inaccessible;
- *Integrity*: received data are not tampered with during transmission; if this does not happen, then any change can be detected;
- *Authenticity*: data senders can always be verified and data receivers cannot be spoofed.

There is an additional property of security that should always be taken into account: authorization. Authorization means that data can be accessed only by those allowed to do so; it should be unavailable to others. This aspect, which requires identification of the communication endpoints, is particularly relevant in those scenarios where it is necessary to ensure that private data cannot be accessed by unknown or unauthorized parties.

It is a common opinion that in the near future IP will be the base common network protocol for the IoT. This does not imply that all objects will be able to run IP; there will always be tiny devices, such as tiny sensors or RFID tags that will be organized in closed networks implementing very simple and application-specific communication protocols and that eventually will be connected to an external network through a suitable gateway. However, it is foreseen that all other small networked objects will exploit the benefits of IP and the corresponding protocol suite.

Bormann has tried to define the classes of constrained devices, in terms of memory capacity, in order to be used as a rough indication of device capabilities [104]:

- Class 1: RAM size = ~10 kB, Flash size = ~100 kB;
- Class 2: RAM size = ~50 kB, Flash size = ~250 kB;

Some of these networked objects, with enough memory, computational power, and power supply, will simply run existing IP-based protocol suite implementations. Others will still run standard Internet protocols, but may benefit from specific implementations that try to achieve better performance in terms of memory size, computational power, and power consumption. In other constrained networked scenarios, smart objects may require additional protocols and some protocol adaptations in order to optimize Internet communications and lower memory, computational, and power requirements.

There is currently considerable effort within the IETF to extend existing protocols for use in resource-constrained networked environments. Some of the current IETF working groups targeted to these environments are:

- Constrained RESTful Environments (CoRE) [21];
- IPv6 over Low Power WPAN (6LoWPAN) [19];
- Routing over Low Power and Lossy Networks (ROLL) [20];
- Lightweight Implementation Guidance (LWIG) [105].

In Figure 5.1, a typical IP-based IoT protocol stack is depicted and compared with the classical Internet protocol stack used by standard non-constrained nodes for accessing the web. At the application layer, the HTTP [2] protocol is replaced by the Constrained Application Protocol (CoAP) [7], which is an application layer protocol to be used by resource-constrained devices. It offers a representational state transfer (REST) service for machine-to-machine (M2M) communications, and can be easily translated to/from HTTP.

Significant reasons for proper protocol optimizations and adaptations for resource-constrained objects can be summarized as follows:

- Smart objects typically use, at the physical and link layers, communication protocols (such as IEEE 802.15.4) that are characterized by small maximum transmission units, thus leading to packet

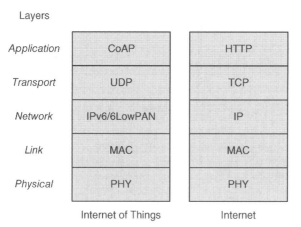

Layers

Application	CoAP	HTTP
Transport	UDP	TCP
Network	IPv6/6LowPAN	IP
Link	MAC	MAC
Physical	PHY	PHY

Internet of Things   Internet

Figure 5.1 Comparison between the IoT and the Internet protocol stack for OSI layers.

fragmentation. In this case, the use of compressed protocols can significantly reduce the need for packet fragmentation and postponed transmissions.

- Processing larger packets likely leads to higher energy consumption, which can be a critical issue in battery-powered devices.
- Minimized versions of protocols (at all layers) can reduce the number of exchanged messages.

Protocol compression is especially relevant when dealing with security protocols, which typically introduce higher overhead and increase the size of transmitted data packets. Besides protocol compression, cross-layer interaction between protocols plays a crucial role. This is particularly important in order to avoid useless duplication of security features, which might have a detrimental impact on the computation and transmission performance. For instance, end-to-end security can be guaranteed by adopting IPSec at the network layer or TLS/DTLS at the transport layer. Combining these two security protocols results in very expensive processing, both at the secure channel setup phase and during packet transmission/reception.

Another important issue in the introduction of security protocols is interoperability. Security protocols typically allow the negotiation of some parameters to be used during operations. Such negotiations might be related to cryptographic and digital signature algorithms. In order to guarantee full interoperability among smart objects, it is necessary to define a set of mandatory options, which all objects must implement for minimal support. The algorithms supported by an object are declared in a negotiation phase and a suitable choice is then agreed upon by the two communicating parties. It is not necessary that the mandatory algorithms are standard algorithms used in the Internet, but can be ones targeted for use in constrained environments.

A final remark should be made about the heterogeneous nature of smart objects, whose characteristics can vary significantly with relevant differences with respect to those of conventional hosts. This means that the adoption of a suite of security protocols and cryptographic algorithms is a need and a challenge at the same time. Standardization can lead to full interoperability, yet it is extremely difficult to agree on a set of protocols and algorithms that will be supported by all devices.

5.2 Security Mechanisms Overview

As mentioned in Section 5.1, one of the most important requirements and crucial aspects for a correct deployment and diffusion of IoT is security. Several challenging security goals must be achieved, including data confidentiality, data authentication, integrity, service availability, peer entity authentication, authorization, anonymity, and/or pseudonymity. Since the protocol architecture of smart objects should adhere to standard IP architecture (for obvious integration reasons), many of the security mechanisms already defined and currently used for the Internet can be reused in IoT scenarios. Moreover, since many Internet security protocols allow for the possibility of selecting and suitably configuring the algorithms and other cryptographic primitives used, they can be reused, although possibly with suitable algorithmic or configuration modifications.

In this section, the main protocols for securing IP-based end-to-end communications between smart objects are reviewed, and the main issues related to this type of communication are discussed. Algorithms and other mechanisms actually used by these protocols are discussed in Section 5.2.2.

5.2.1 Traditional vs Lightweight security

According to the protocol stacks depicted in Figure 5.1, a direct comparison between possible layered architectures of security protocols in Internet and IoT scenarios is shown in Figure 5.2.

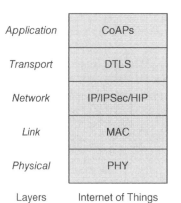

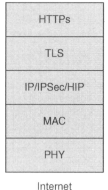

Figure 5.2 Comparison of Internet and IoT security protocols.

Layers	Internet of Things	Internet
Application	CoAPs	HTTPs
Transport	DTLS	TLS
Network	IP/IPSec/HIP	IP/IPSec/HIP
Link	MAC	MAC
Physical	PHY	PHY

The IoT protocol suite depicted in Figure 5.2 represents only the possible choices for a smart object to enforce data protection (at different layers), rather than the actual set of security mechanisms effectively implemented and simultaneously used at different layers. However, in order to minimize the used resources, particular attention has to be devoted to avoid the repetition of the same functionalities at different layers, if not strictly required.

Referring to the IoT protocol stack of Figure 5.2, at the application layer is the CoAP application protocol, which can be used for request/response interactions between smart objects or between a smart object and a non-constrained (standard) Internet node (possibly by using some intermediate relay/proxy node). CoAP itself does not provide primitives for authentication and data protection, so these functions should be implemented directly at the application/service layer (by directly protecting the data encapsulated and exchanged by CoAP) or at one of the underlying layers. Although data authentication, integrity, and confidentiality can be provided at lower layers, such as PHY or MAC (e.g., in IEEE 802.15.4 systems), no end-to-end security can be guaranteed without a high level of trust on intermediate nodes. However, due to the highly dynamic nature of the wireless multi-hop communications expected to be used to form the routing path between remote end nodes, this kind of security (hop-by-hop) is not, in general, sufficient. For this reason, security mechanisms at network, transport, or application levels should be considered instead of (or in addition to) PHY- and MAC-level mechanisms.

5.2.1.1 Network Layer

At the network layer, an IoT node can secure data exchange in a standard way by using Internet Protocol Security (IPsec) [106]. IPSec was originally developed for IPv6, but found widespread deployment, first, as an extension of IPv4, into which it was back-engineered. IPSec was an integral part of the base IPv6 protocol suite, but has since then been made optional. IPSec can be used in protecting data flows between a pair of hosts (host-to-host communication), between a pair of security gateways (network-to-network communication), or between a security gateway and a host (network-to-host communication).

For each IP packet, IPSec can provide confidentiality, integrity, data-origin authentication and protection against replay attacks (it works at the network layer). Such security services are implemented by two IPSec security protocols: Authentication Header (AH) and Encapsulated Security Payload (ESP). While AH provides integrity, data-origin authentication, and optionally anti-replay capabilities, ESP can provide confidentiality, data-origin authentication, integrity, and anti-replay capabilities.

IPSec AH and ESP define only the way payload data (clear or enciphered) and IPSec control information are encapsulated, while the effective algorithms for data origin authentication/integrity/confidentiality are specified separately and selected from amongst a set of available cipher suites. This modularity makes IPSec usable in the presence of very resource-constrained devices, if a suitable algorithm that guarantees both usability and adequate security is selected. This means that, from an algorithmic point of view, the problem moves from the IPSec protocol itself to the actual cryptographic algorithms. Section 5.2.2 is dedicated to algorithm-related issues.

The keying material and the selected cryptographic algorithms used by IPSec for securing a communication are called an IPSec Security Association (SA). To establish an SA, IPSec can be pre-configured (specifying a pre-shared key, hash function and encryption algorithm) or can be dynamically negotiated by the IPSec Internet Key Exchange (IKE) protocol. Unfortunately, as the IKE protocol was designed for standard Internet nodes, it uses asymmetric cryptography, which is computationally heavy for very small devices. For this reason, suitable IKE extensions using lighter algorithms should be considered. These issues are considered in Section 5.2.2.

Other problems related to the implementation of IPSec in constrained IoT nodes include data overhead (with respect to IP), configuration, and practical implementation aspects. Data overhead is introduced by the extra header encapsulation of IPSec AH and/or ESP. However, this can be limited by implementing header compression techniques, similar to what is done in 6LoWPAN for the IP header. A possible compression mechanism for IPSec in 6LoWPAN has been proposed and numerically evaluated [107].

Regarding the practical aspects, it is worth noting that IPSec is often designed for VPNs, thus making it difficult for them to be dynamically configurable by an application. Moreover, existing implementations are also barely compatible with each other and often require manual configuration to interoperate.

An alternative to using IKE+IPsec is the Host Identity Protocol (HIP) [108]. The main objective of HIP is to decouple the two functions of host locator (for routing purposes) and host identifier (for actual host identification) currently performed by IP addresses. For this purpose, HIP introduces a new namespace between IP and upper layers, specific to host identification and based on public cryptography. In HIP, the host identity (HI) is directly associated with a pair of public/private keys, where the private key is owned by the host and the public key is used as the host identifier. HIP defines also an host identity tag (HIT), a 128-bit representation of the HI based on the hash of the HI plus other information. This can be used, for example, as a unique host identifier in the existing IPv6 API and by application protocols. HIP also defines an HIP exchange that can be used between IP hosts to establish a HIP security association, which in turn can be used to start secure host-to-host communications based on the IPSec ESP protocol [109].

In addition to security, HIP provides methods for IP multi-homing and host mobility, which are important features for an IP-based IoT network architecture. Some work is also being carried out to let the HIP exchange run on very constrained devices. The approach involves using suitable public-key cryptographic primitives, such as the ones described in Section 5.2.2.

5.2.1.2 Transport Layer

In the current IP architecture, data exchange between application nodes can be secured at the transport layer through standard Transport Layer Security (TLS) and Datagram Transport Layer Security (DTLS) protocols. TLS is the most common secure protocol, running on top of the TCP, and providing to the application layer the same connection and stream-oriented interface as TCP [110]. In addition, TLS provides complete secure communication through:

- peer-entity authentication and key exchange (using asymmetric cryptography);
- data authentication, integrity, and anti-replay (through message authentication code);
- confidentiality (using symmetric encryption).

Peer-entity authentication and key exchange is provided by the TLS handshake phase, which is performed at the beginning of the communication.

DTLS, on the other hand, was introduced more recently, in order to provide a security service similar to TLS on top of UDP [111].

Although it is still poorly supported in standard Internet nodes, it is currently the reference security protocol for IoT systems since it uses UDP as transport and does not suffer from the problems caused by the use of TCP in network-constrained scenarios (due to the extremely variable transmission delay and lossy links).

Both IPSec and DTLS provide the same security features, but with their own mechanisms and at different stack layers. Moreover, the IPSec IKE key agreement is almost the same as the DTLS handshake function. The main advantage of securing communications at the transport layer with DTLS consists in allowing more precise access control. In fact, operation at the transport layer allows applications to directly and easily select which, if any, security service has to be set up. Another practical advantage is that DTLS allows for the reuse of the wide experience gained during implementations of TLS.

For these reasons DTLS has recently received significant attention as a possible way of securing communication of constrained node/network applications and it has been standardized as the security protocol for CoAP as associated to coaps URIs [7].

Unfortunately, there are still some few issues that must be faced in order to make DTLS more friendly for constrained devices. The most important are related to the limited packet size imposed by the underlying protocols, such as IEEE 802.15.4. In fact, as for IPSec, DTLS introduces an overhead during both handshake and data transport phases. DTLS causes fragmentation in the handshake layer, and this can add a significant overhead. Another solution might be to use the fragmentation offered at IPv6 or the 6LoWPAN layer. Moreover, in order to reduce DTLS overhead, a packet optimization and compression mechanism can be introduced. For example Raza *et al.* proposed using the 6LoWPAN compression mechanisms for the DTLS protocol [112].

From the security point of view, one problem of using DTLS or IPSec is that end-to-end communication is not guaranteed when intermediate nodes such as proxies or application-level gateways are introduced. In fact, both IPSec and DTLS provide secure communications at IP and transport layers respectively, and, in the presence of multi-hop application-level communications, they can ensure security only within each hop. In addition, some complications in providing end-to-end security may also arise when connectivity is realized directly at IP and transport layers. There are scenarios in which a part of a network (internal) of constrained devices is interconnected at IP level to the rest of an (external) network, for example the Internet. Although data protection can be guaranteed through IPSec or DTLS protocols, other network attacks, like flooding or replay, may occur due to the asymmetry of the resources available at the end systems; for example, a high-powered host attached to the Internet may attack a constrained device by trying to consume all of its limited power or processing resources. In order to guarantee a suitable level of protection against this kind of attack, an intermediate security gateway may be required at the border of the internal network. A security gateway may act as access controller, granting access to the internal network only to trusted nodes. Solutions to this problem have been proposed [113, 114]. In particular, in the case of end-to-end application-level communication based on CoAP, a solution may be to require the external node to encapsulate CoAP/DTLS/IP traffic within a proper DTLS tunnel established between the external node and the security gateway.

It is also important to note that, although DTLS provides a datagram-oriented communication service (like UDP), it establishes a point-to-point secure association that is not compatible with multicast communications (in contrast to UDP, which does support multicast). In order to make DTLS applicable in multicast IP-communication scenarios, some protocol extensions for group-key management will be needed in the future.

5.2.1.3 Application Layer

Providing security at the IP layer (through IPSec) or the transport layer (through TLS or DTLS) has several advantages. The main ones are:

- The same standard mechanism and the same implementation can be shared by all applications, resulting in code reuse and reduced code size.
- Programmers do not have to deal with the implementation of any security mechanism; this significantly simplifies the development of applications when secure communications are required.

Unfortunately, as already described, both IPSec and (D)TLS have their own drawbacks. Probably the main one, common to both IP and transport approaches, is the impossibility to ensure complete end-to-end security when application communications are relayed by intermediate nodes that work at application level (e.g., proxies). In this case, end-to-end security can be still provided with transport- or IP-level mechanisms, but only in the presence of very trusted intermediate systems. However, in this case, the overall security is complicated by the handling of such hop-by-hop trust management.

A different approach to providing complete end-to-end security is to enforce security directly at the application level. This of course simplifies the requirements for underlying layers, and probably reduces the cost, in term of packet size and data processing, since only application data have to be secured, and only per-data and not per-packet overhead is introduced. Moreover, multicast communication and in-network data aggregation in encrypted domains (for example through homomorphic cryptography) are easier to implement at application level.

The main disadvantages of providing security at application level are the complications introduced for application development and the overall code size caused by poor reuse of software code. This is mainly due to the lack of well-defined and adopted secure protocols at application level. Examples of standards that can be used for this purpose are S/MIME and SRTP. S/MIME (Secure/Multipurpose Internet Mail Extensions) [115] is a standard for providing authentication, message integrity, non-repudiation of origin, and confidentiality for application data. Although S/MIME was originally developed for securing MIME data between mail user agents, it is not restricted to mail and can be used for securing any application data and can be encapsulated within any application and transport protocol.

SRTP (Secure Real-time Transport Protocol) [116] is another secure communication protocol that provides confidentiality, message authentication, and replay protection to application data. It is an extension of the Real-time Transport Protocol (RTP) specifically developed for handling real-time data communications (e.g., voice or video communication), but can also be re-used in other application scenarios. It works in a per-packet fashion and is usually encapsulated in UDP.

More investigation is required to state which is the standard protocol most suitable for securing data at application layer in network- and node-constrained scenarios such as the IoT.

5.2.2 Lightweight Cryptography

The development of the IoT will result in the deployment of billions of smart objects that will interact with the existing Internet. Smart objects are tiny computing devices, with constrained resources: low computation capabilities, little memory, and limited battery lives. Communication with smart objects in resource-constrained environments must necessarily take into account these hard limitations, especially in scenarios where security is crucial and conventional cryptographic primitives, such as the Advanced Encryption Standard (AES) [117], are inadequate.

Lightweight cryptography (LWC) is a very interesting research area, aiming at the design of new ciphers that might meet the requirements of smart objects [118]. The term "lightweight" should not be mistaken as meaning "weak" (in terms of cryptographic protection), but should instead be interpreted as referring to a family of cryptographic algorithms with smaller footprint, lower energy consumption, and low computational power needs. These ciphers aim at providing sufficient security in the environment of restricted resources that is encountered in many ubiquitous devices [119]. LWC thus represents a cryptography tailored to constrained devices, which must cope with the trade-offs between security level, cost, and performance.

In this section, an overview of the most prominent cryptographic algorithms is presented, followed by a comparison of lightweight cryptographic primitives and conventional ones, such as AES, which are currently adopted in standard Internet security protocols, such as IPSec and TLS. Symmetric ciphers for lightweight cryptography are presented first, followed by asymmetric ciphers and then cryptographic hash functions. Finally, privacy homomorphism is discussed. We note that this overview is not meant to be detailed or extensive, but aims at pointing out which encryption algorithms are most suitable for practical implementation in IoT scenarios.

5.2.2.1 Symmetric-key LWC Algorithms

Symmetric-key cryptographic algorithms use the same key for encryption of a plaintext and decryption of a ciphertext. The encryption key represents a shared secret between the parties that are involved in the secure communication. An illustrative representation of symmetric-key secure communication is shown in Figure 5.3.

Symmetric-key encryption can use either block ciphers or stream ciphers:

- *Block ciphers* operate on fixed-length groups of bits, called blocks, padding the plaintext to make its length equal to a multiple of the block size. An example is the AES algorithm.

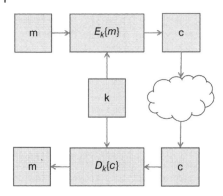

Figure 5.3 Secure communication with symmetric-key cryptographic algorithms.

- In *stream ciphers* the digits of a plaintext are encrypted one at a time with the corresponding digit of a pseudorandom cipher digit stream (keystream).

Tiny Encryption Algorithm

The Tiny Encryption Algorithm (TEA) is a block cipher renowned for its simplicity of description and implementation; typically a few lines of code [120]. TEA operates on two 32-bit unsigned integers (which could be derived from a 64-bit data block) and uses a 128-bit key. TEA relies only on arithmetic operations on 32-bit words and uses only addition, XORing, and shifts. TEA uses a large number of iterations, rather than a complicated program, in order to avoid preset tables and long setup times. The main design goal of TEA is to define a simple and short cipher that does not rely on preset tables or pre-computations, thus leading to a smaller footprint.

TEA has been revised in order to fix some weaknesses found in the original algorithm, such as the problem of equivalent keys, which reduced the actual key size from 128 to 126 bits. The redesign of TEA, named XTEA (extended TEA) [121], fixes this problem by changing the key schedule. XTEA also requires two fewer additions, thus resulting in a slightly faster algorithm. Other modifications of the TEA algorithm have been presented, such as XXTEA, block TEA, speed TEA, and tiny XTEA.

As the TEA family uses exclusively very simple operations and has a very small code size, it is an ideal candidate as a cryptographic algorithm for implementing security mechanisms in smart objects and wireless sensors.

Scalable Encryption Algorithm

The Scalable Encryption Algorithm (SEA) is targeted at small embedded applications [122]. The design considers a context with very limited processing resources and throughput requirements. Another design principle of SEA is flexibility: the plaintext size n, key size n, and processor (or word) size b are design parameters, with the only constraint that n is a multiple of $6b$; for this reason, the algorithm is denoted as $SEA_{n,b}$. The motivation of this flexibility is the observation that many encryption algorithms perform differently depending on the platform, e.g., 8-bit or 32-bit processors. $SEA_{n,b}$ is designed to be generic and adaptable to different security levels (by varying the key size) and target hardware. A great advantage of $SEA_{n,b}$ is the "on-the-fly" key derivation. The main disadvantage is that $SEA_{n,b}$ trades space for time and this may not be negligible on devices with limited computational power.

PRESENT Cipher

PRESENT is an ultra-lightweight block cipher algorithm based on a substitution-permutation network [123]. PRESENT has been designed to be extremely compact and efficient in hardware. It operates on 64-bit blocks and with keys of either 80 or 128 bits. It is intended to be used in situations where low-power consumption and high chip efficiency are desired, thus making it of particular interest for constrained environments. The main design goal of PRESENT is, as for the other lightweight ciphers, simplicity. PRESENT is performed in 31 rounds, each comprising three stages:

- key-mixing, through XOR operation and a 61-bit rotation key schedule;
- substitution layer, through 16 4-bit (input) by 4-bit (output) S-boxes;
- permutation layer.

At the end of the 31st round, an additional round is performed by XORing the last-round subkey.

ISO/IEC 29192-2:2012 *Lightweight Cryptography* names PRESENT as a block cipher suitable for lightweight cryptography [124].

HIGHT

The HIGh security and lightweigHT (HIGHT) encryption algorithm is a generalized Feistel network with a block size of 64 bits, 128-bit keys and 32 rounds [125]. HIGHT was designed with an eye on low-resource hardware performance. HIGHT uses very simple operations, such as XORing, addition mod 2^8, and bitwise rotation. The key schedule in HIGHT is designed so that subkeys are generated on the fly both in the encryption and the decryption phases.

Comparison of Symmetric LWC Algorithms

LWC algorithms are not intended to supersede existing ciphers, such as AES, for widespread use. Their application is limited to those scenarios where classical ciphers might be inefficient, such as scenarios where:

- a moderate security level is required, so that keys need not be too long;
- encryption should not be applied to large amounts of data;
- the hardware area needed for implementation and the power consumption are considered harder requirements than speed.

For constrained devices, the choice of the cryptographic algorithm is a primary element that can affect performance. When low cost and energy consumption are hard requirements, computational power must inherently be downsized accordingly. Using 8-bit microcontrollers (such as Atmel AVR microcontrollers [126]), which have limited capabilities in terms of computing power, memory, and storage, requires that implemented ciphers have small footprints and are kept simple. This may result in faster execution and thus in lower energy consumption, which may be critical for battery-powered devices.

Although most symmetric cryptographic algorithms have been developed with a focus on efficient software implementations, the deployment of smart objects will lead to an increasing attention being given to those ciphers that will perform well in hardware in terms of speed and energy consumption. In Table 5.1, we report a direct comparison of the LWC algorithms outlined in Subsection 5.2.2.1 [118], with particular reference to the following metrics: key size, block size, rounds, consumed area measured in gate equivalents (GEs), and code size (in bytes). Reported values for gate equivalents are related to hardware implementations, while code size refers to software implementations.

5.2.2.2 Public-key (Asymmetric) LWC Algorithms

Public-key (asymmetric) cryptography requires the use of a public key and a private key. Public keys can be associated with the identity of a

Table 5.1 Comparison of different symmetric-key cryptographic algorithms.

	Cipher	Key size (bits)	Block size (bits)	Rounds (hardware impl.)	GE (software impl.)	Code size (bytes)
Software ciphers	AES	128	128	10	3400 [127, 128]	2606
	TEA	128	64	32	3490 [129]	1140
	$SEA_{96,8}$	96	8	$\geq 3n/4$	3758 [a] [130] 3925 [b] [130] 2547 [131]	2132
Hardware ciphers	PRESENT	80	64	32	1570 [123]	936
	HIGHT	128	64	32	3048 [125]	5672

a) Round-based implementation with datapath of size n
b) Serialized implementation with datapath of size b.

node by including them in a public certificate, signed by a certification authority, which can be asked to verify the certificate. Public-key cryptography requires a significant effort to deploy a public-key infrastructure. Moreover, asymmetric cryptography requires higher processing and long keys (at least 1024 bits for RSA [132]). Alternative public-key cryptographic schemes, such as elliptic curve cryptography [133], might require shorter keys to be used in order to achieve the same security as RSA keys. However, because of this, symmetric cryptography is preferred in terms of processing speed, computational effort, and size of transmitted messages. Public-key ciphers are usually used to set up symmetric keys to be used in subsequent communications.

RSA Algorithm

The Rivest, Shamir, and Adleman (RSA) algorithm is the best known and widely used public-key scheme. It is based on exponentiation in a finite field over integers modulo N. Consider a modulus N and a pair of public and private keys (e, d). The encryption of a message m is given by $c = m^e \bmod N$, while the decryption is $m = c^d \bmod N$. The key generation phase of RSA, aiming to generate the public–private key pair, consists of the following steps:

1) Select two large prime numbers denoted as p and q such that $p \neq q$.
2) Compute $n = p \cdot q$.
3) Compute the Euler's totient function $\Phi(n) = (p - 1) \cdot (q - 1)$.
4) Choose an integer e such that $1 < e < \Phi(n)$ and that the $GCD(e, \Phi(n)) = 1$.
5) Compute $d = e^{-1} \bmod \Phi(n)$.

The pair (n, e) is the public key, while d is the private key.

The security of the RSA algorithm depends on the hard problem of factorizing large integers. In order to achieve an acceptable level of security, n should be at least 1024 bits long, so that p and q, and consequently $\Phi(n)$, cannot be obtained, thus protecting the (e, d) pair.

RSA is unsuitable for adoption in constrained devices due to the need to operate on large numbers and the fact that long keys are required to achieve sufficient security. Moreover, both key generation and encryption/decryption are demanding procedures that result in higher energy consumption.

Elliptic Curve Cryptography

Elliptic curve cryptography (ECC) is an approach to public-key cryptography based on the algebraic structure of elliptic curves over finite fields. While RSA is based on exponentiation on finite fields, ECC depends on point multiplication on elliptic curves. An elliptic curve E over the finite field K (whose characteristic is not equal to 2 and 3) is defined as:

$$E(K) : y^2 = x^3 + ax + b \text{ with } a, b \in K$$

Points $P = (x, y) \in E(K)$ form an Abelian group, so point addition and scalar point multiplication can be performed.

ECC provides higher security and a better performance than the first-generation public-key techniques, RSA and Diffie–Hellman. Moreover, ECC is the most interesting public-key cryptographic family for embedded environments because it can reach the same security level as RSA with much shorter keys, as shown in Table 5.2, and with computationally lighter operations, like addition and multiplication, rather than exponentiation.

ECC has been accepted commercially and has also been adopted by standards institutions such as the American National Standards Institute (ANSI), the Institute of Electrical and Electronics Engineers (IEEE), the International Organization for Standardization (ISO), the Standards for Efficient Cryptography Group (SECG), and the National Institute of Standards and Technology (NIST) [134–138].

The implementation of a lightweight hardware ECC processor for constrained devices is attracting growing interest. A possible hardware implementation of a low-area, standalone, public-key engine for ECC, with a 113-bit binary field for short-term security and a 193-bit

Table 5.2 Comparison of security levels for symmetric ciphers, ECC, and RSA (recommended NIST key sizes).

Symmetric key size (bits)	80	112	128	192	256
ECC key size (bits)	160	224	256	384	512
RSA key size (bits)	1024	2048	3072	7680	15360

Source: http://www.nsa.gov/business/programs/elliptic_curve.shtml

binary field for medium-term security, has been demonstrated [118]. The choice of a binary field, rather than a prime field, is related to the corresponding carry-free arithmetic, which fits well in hardware implementations. With respect to other ECC hardware implementations, the one presented uses a smaller area (in terms of GEs) and exhibits faster execution.

Performance Comparison of Public-key Cryptographic Algorithms

Here we review the performance results [139] for implementations of RSA and ECC public-key algorithms, such as TinyECC and Wiselib, against benchmarks obtained in constrained devices (namely an 8-bit Arduino Uno board). Table 5.4 shows the implementation results for RSA public-key encryption, when the private key is held in SRAM or in ROM. In Table 5.5, the performance of ECDSA signature algorithms in TinyECC and Wiselib implementations is compared. A comparison of the ROM footprints is shown in Table 5.3.

5.2.2.3 Lightweight Cryptographic Hash Functions

Cryptographic hash functions, such as MD5 [140] and SHA-1 [141], are an essential part of any protocol that uses cryptography. Hash functions are used for different purposes, such as message integrity check,

Table 5.3 Public-key encryption library ROM occupancy.

Library	AvrCryptolib	Wiselib	TinyECC	Relic-toolkit
ROM footprint (kB)	3.6	16	18	29

Table 5.4 RSA private key operation performance.

Key length (bits)	Execution time (ms)		Memory footprint (bytes)	
	Key in SRAM	Key in ROM	Key in SRAM	Key in ROM
64	66	70	40	32
128	124	459	80	64
512	25089	27348	320	256
1024	109666	218367	640	512
2048	1587559	1740267	1280	104

Table 5.5 ECDSA signature performance: TinyECC versus Wiselib implementations.

Curve parameters	Execution time (ms)		Memory footprint (bytes)		Comparable RSA key length
	TinyECC	Wiselib	TinyECC	Wiselib	
128r1	1858	10774	776	732	704
128r2	2002	10615	776	732	704
160k1	2228	20164	892	842	1024
160r1	2250	20231	892	842	1024
160r2	2467	20231	892	842	1024
192k1	3425	34486	1008	952	1536
192r1	3578	34558	1008	952	1536

digital signatures, and fingerprinting. Cryptographic hash functions should ideally be:

- computationally inexpensive;
- pre-image resistant: given a hash h, it should be difficult to invert the hash function in order to obtain the message m such that $h = hash(m)$;
- second pre-image resistant: given a message m_1, it should be difficult to find another message m_2 such that $hash(m_1) = hash(m_2)$;
- collision resistant: it should be difficult to find two messages m_1 and m_2, with $m_1 \neq m_2$, such that $hash(m_1) = hash(m_2)$ (hash collision).

In general, for a hash function with n-bit output, pre-image and second pre-image resistance require 2^n operations, while collision resistance requires $2^{n/2}$ operations [142]. While the design of standard cryptographic hash functions does not focus on hardware efficiency, lightweight cryptographic hash functions are needed for use in resource-constrained devices in order to minimize the amount of hardware (in terms of GEs) and energy consumption.

In this subsection, we will overview some proposals for lightweight cryptographic hash functions that go beyond the classical MD and SHA families.

DM-PRESENT and H-PRESENT

Bogdanov *et al.* have proposed DM-PRESENT and H-PRESENT, two lightweight hash functions based on the PRESENT block cipher [142].

DM-PRESENT is a 64-bit hash function and comes in two versions: DM-PRESENT-80 and DM-PRESENT-128, depending on which cipher (PRESENT-80 or PRESENT-128) is used. H-PRESENT (namely H-PRESENT-128) is a 128-bit hash function based on the PRESENT-128 block cipher. In their work, the authors also considered the problem of constructing longer hash functions based on the PRESENT block cipher in order to improve the security level.

PHOTON

PHOTON [143] is a hardware-oriented family of cryptographic hash functions designed for constrained devices. PHOTON uses a sponge-like construction [144] as domain extension algorithm and an AES-like primitive as an internal unkeyed permutation. A PHOTON instance is defined by its output size ($64 \leq n \leq 256$), its input rate r, and its ouptut rate r' (PHOTON-$n/r/r'$). The use of a sponge function framework aims at keeping the internal memory usage low. The framework has been extended in order to increase speed when hashing small messages, which is typically inefficient in a sponge function framework.

SPONGENT

SPONGENT [145] is a family of lightweight hash functions with outputs of 88, 128, 160, 224, and 256 bits. SPONGENT is based on a sponge construction with a PRESENT-type permutation. An instance of SPONGENT is defined by the output size n, the rate r, and the capacity c (SPONGENT-$n/c/r$). The size of the internal state, denoted as width, is $b = r + c \geq n$. Implementations in ASIC hardware require 738, 1060, 1329, 1728, and 1950 GEs, respectively, making it the hash function with the smallest footprint in hardware. The 88-bit hash size is used only to achieve pre-image resistance.

QUARK

The QUARK [146] hash family comes with three instances: U-QUARK, D-QUARK, and S-QUARK, with hash sizes of 136, 176, and 256 bits, respectively. QUARK, like PHOTON and SPONGENT, is based on a sponge construction. The QUARK hash family is optimized for hardware implementation and, as stated by the authors, software implementations should instead rely on other designs. QUARK has a bigger footprint than PHOTON and SPONGENT, but shows higher throughput than SPONGENT and better security than PHOTON.

Keccak

Keccak [147] is a family of sponge functions. Keccak uses a sponge construction in which message blocks are XORed into the initial bits of the state, which is then invertibly permuted. In the version used in Keccak, the state consists of a 5×5 array of 64-bit words: 1600 bits in total. Keccak produces an arbitrary output length.

Keccak was selected by the NIST as the winner of the SHA-3 competition [148] on October 2, 2012. Since that time it has been referred to as SHA-3.

SQUASH

SQUASH (SQUare-hASH) [149] is suited to challenge-response MAC applications in constrained devices, such as RFID tags. SQUASH is completely deterministic, so it requires no internal source of randomness. SQUASH offers 64-bit pre-image resistance. SQUASH is not collision resistant, but this is not an issue since it targets RFID authentication protocols, where collision resistance is not needed. If collision resistance is a requirement, for instance for digital signatures, SQUASH is unsuitable and other hash functions should be considered.

5.2.2.4 Homomorphic Encryption Schemes

Homomorphic encryption is a form of encryption that allows specific types of computation to be executed on ciphertexts to give an encrypted result that is the ciphertext of the result of operations performed on the plaintext. By denoting $E\{\cdot\}$ as the homomorphic encryption function and $f(\cdot)$ as the computation function, it holds that:

$$E\{f(a, b)\} = f(E\{a\}, E\{b\})$$

An example of homomorphic encryption is the RSA algorithm. Consider a modulus N and an exponent e. The encryption of a message m is given by $E\{m\} = m^e \bmod N$. The homomorphic property holds, since:

$$E\{m_1 \cdot m_2\} = (m_1 \cdot m_2)^e$$
$$\bmod N = (m_1)^e \bmod N \cdot (m_2)^e \bmod N = E\{m_1\} \cdot E\{m_2\}$$

Other examples of homomorphic encryption schemes are the ECC encryption [133], the ElGamal cryptosystem [150] and the Pailler cryptosystem [151].

Homomorphic encryption is receiving a growing interest for application in IoT scenarios, since it could be used to preserve confidentiality among the endpoints of communication, while making it possible for intermediate nodes to process information without the need to decrypt the data prior to processing. Homomorphic cryptosystems usually require higher levels of computation and need longer keys to achieve a comparable security level than symmetric-key algorithms.

Depending on the operation $f(\cdot)$ that can be performed on the encrypted data, the homomorphic encryption scheme can be defined as additive or multiplicative. Additive homomorphism makes it possible to compute sums, subtractions, and scalar multiplication of its operands; multiplicative homomorphism allows computation of the product of its operands. The RSA algorithm is an example of multiplicative homomorphic encryption. An example of additive homomorphic encryption is the Pailler cryptosystem. Given a modulus n, a shared random integer g, and user-generated random integers r_1 and r_2, the homomorphic property is:

$$E\{m_1\} \cdot E\{m_2\} = (g^{m_1} r_1^n \bmod n^2) \cdot (g^{m_2} r_2^n \bmod n^2)$$
$$= (g^{m_1+m_2})(r_1 r_2)^n \bmod n^2 = E\{m_1 + m_2\}$$

Homomorphic encryption schemes that are either additive or multiplicative are termed "partially homomorphic". If both addition and multiplication are supported, a cryptosystem is called "fully homomorphic". Fully homomorphic cryptosystems preserve the ring structure of the plaintexts and, therefore, enable more complex procedures to be used. The investigation of fully homomorphic encryption schemes is still in its early stages and no practical scheme with acceptable performance has been found (e.g., in terms of decryption delay). Application of these schemes to IoT scenarios is a rich research topic.

5.2.3 Key Agreement, Distribution, and Security Bootstrapping

Key distribution and management is a crucial issue that needs to be addressed when security mechanisms have to be adopted. Key agreement protocols have been around for years: the Diffie–Hellman key exchange protocol is an example of a key agreement protocol that two parties perform in order to setup a shared key to be used in a session [152]. Other mechanisms have been defined and implemented.

The Internet Key Exchange (IKE) [153] protocol is a the protocol defined to setup a secure association to be used in IPSec.

5.2.3.1 Key Agreement Protocols

Asymmetric (public-key) cryptographic algorithms are often the basis for key agreement protocols, although other techniques that do not involve the adoption of asymmetric cryptography have been proposed. A polynomial-based key pre-distribution protocol has been defined [154] and applied to wireless sensor networks (WSNs) [155]. A possible alternative key agreement protocol is SPINS [156], which is a security architecture specifically designed for sensor networks. In SPINS, each sensor node shares a secret key with a base station, which is used as a trusted third-party to set up a new key, thus avoiding use of public-key cryptography. Chan *et al.* have presented three efficient random key pre-distribution schemes for solving the security bootstrapping problem in resource-constrained sensor networks, each of which represents a different tradeoff in the design space of random key protocols [157].

5.2.3.2 Shared Group-key Distribution

The mechanisms just described apply to scenarios in which communication occurs between two parties (unicast and point-to-point communications). In other communication scenarios, such as point-to-multipoint (multicast) or multipoint-to-point communications, other mechanisms must be used. In such scenarios, the adoption of a shared group key is appealing.

Secure group communications ensure confidentiality, authenticity, and integrity of messages exchanged within a group through the use of suitable cryptographic services and without interfering with the communication data path. In order to achieve secure group communication, nodes must share some cryptographic material that must be handled in a way that allows any group membership changes, both predictable and unpredictable, to be managed. In fact, any membership change should trigger a rekeying operation, which updates and redistributes the cryptographic material to the group members. This ensures that:

- a former member of the groups cannot access current communications ("forward secrecy") [158];
- a new member cannot access previous communication ("backward secrecy" [159]).

Keoh [160] define an approach, based on DTLS records, to secure multicast communication in lossy low-power networks.

Assuming that the cryptographic primitives used cannot be broken by an attacker with limited computational power (i.e., for whom it is infeasible to carry out a brute force attack in order to solve the problems behind cryptographic schemes, such as discrete logarithms, or inverting MD5/SHA-1), the main challenge is the distribution of the group keys and their updates: this problem is referred to as group-key distribution, and can be tackled according to two different approaches:

- current communications can be deciphered independently of previous communications (stateless receivers): this approach is called "broadcast encryption" [161, 162];
- users maintain state of the past cryptographic material (stateful receivers): this approach is termed "multicast key distribution" [163].

In multicast key distribution, centralized [164] or distributed [165] approaches can be adopted. In the distributed approach, the group key is computed and maintained by the group members themselves. An example of a distributed approach is the Tree-based Group Diffie–Hellman protocol [166]. In a centralized approach, the task of key distribution is assigned to a single entity, called the key distribution center (KDC). This approach gives a simple mechanism with a minimal number of exchanged messages. Logical key hierarchy (LKH) [158] and MARKS [167] are key distribution protocols that try to optimize the number of exchanged messages between a KDC and the group members. LKH is based on key graphs, where keys are arranged into a hierarchy and the KDC maintains all the keys. MARKS is a scalable approach and does not need any update message when members join or leave the group predictably. However, MARKS does not address the issue of member eviction and subsequent key revocation.

5.2.3.3 Security Bootstrapping

All key agreement protocols require that some credentials, either public/private key pairs, symmetric keys, certificates, or others, have been installed and configured on nodes beforehand, so that the key agreement procedure can occur securely. Bootstrapping refers to the processing operations required before the network can operate: this requires that proper setup, ranging from link layer to application layer information, must take place on the nodes. Bootstrapping is a very important phase in the lifecycle of smart objects and can affect the way they behave in operational conditions. Even though the bootstrapping phase is outside the scope of this chapter, it is important to consider security bootstrapping mechanisms and architectures [168], so that possible threats, such as cloning or malicious substitution of objects, can be tackled properly. Jennings provides a sketch of a possible protocol to allow constrained devices to securely bootstrap into a system that uses them [169].

5.2.4 Processing Data in the Encrypted Domain: Secure Data Aggregation

In-network data aggregation in WSNs involves executing certain operations (such as sums and averages) at intermediate nodes in order to minimize the number of transmitted messages and the processing load at intermediate nodes, so that only significant information is passed along in the network. This leads to several benefits, such as energy savings, which are crucial for constrained environments, such as low-power and lossy networks. Data aggregation refers to a multipoint-to-point communication scenario that requires intermediate nodes to operate on received data and forward the output of a suitable function applied to such input data. In such scenarios, where privacy of transmitted data is an issue, it might be necessary to send encrypted data. Encryption can be adopted not only to achieve confidentiality, but also to verify the authenticity and integrity of messages.

While secure data aggregation is certainly also an application-related issue in WSNs, optimized communication can also have other positive impacts in some IoT scenarios. For example, in smart parking or critical infrastructure scenarios, there could be benefits from minimizing transmitted data, possibly by adopting the privacy homomorphism algorithms discussed in Section 5.2.2.

A simple aggregation strategy is to queue the payloads of the received packets and send out only one packet with all the information. This approach can bring only limited gains, since only the payloads are considered. Another, more efficient, approach can be used if the aggregator is aware of the type of operation that the final recipient is willing to perform. Consider a scenario where a node is interested in counting all the nodes in the network. Nodes send a

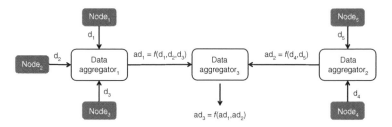

Figure 5.4 In-network data aggregation.

packet with "1" as the content. Aggregators receive these packets and can just sum the 1s received and send out one packet of the same size, whose content is the number of 1s received. By doing this, the information sent across the network is minimal and the final recipient only performs simple processing.

Typically, secure data aggregation mechanisms require nodes to perform the following operations:

1) At the transmitting node, prior to transmission, data are encrypted with some cryptographic function E.
2) At the receiving node, all received data packets are decrypted with the inverse cryptographic function $D = E^{-1}$ to retrieve the original data.
3) Data are aggregated with an aggregation function.
4) Prior to retransmission, aggregated data are encrypted through E and relayed to the next hop.

This process is iterated at intermediate nodes until the data reach the destination node that is interested in receiving the result of aggregation, as shown in Figure 5.4. Both symmetric and asymmetric cryptographic schemes can be applied.

The aggregation procedure just outlined raises the following issues, especially if we consider a scenario where the aggregators are not special nodes, but have the same features as other nodes in the network:

- Aggregators must decrypt each incoming piece of information before processing in order to perform the aggregation and, subsequently, encrypt the result before transmission to the next hop. This has clearly an impact on the computation and, therefore, on the energy consumption of the aggregator.
- An aggregator must keep a secure association (i.e., share a symmetric key) with any node that either sends data to or receives data

from it. This further introduces the need for increased complexity at the aggregator.

- Intermediate aggregators access the data that they receive, even though they are not intended to do so, since the actual recipient of the data is another node. This might introduce privacy concerns, especially in scenarios where intermediate nodes might not have a trust relationship with the sender.

In order to cope with the problems sketched above, various actions can be considered. All these issues can be addressed by using homomorphic encryption schemes, as introduced in Section 5.2.2.4. Homomorphic encryption can be used to avoid the need to decrypt the information that must be aggregated and then encrypt the result; it is possible to operate on the encrypted data directly. This can dramatically increase the performance, in terms of execution time and energy savings, since the encryption/decryption operations are typically computationally demanding. Besides computational and energy efficiencies, a positive side effect of the adoption of homomorphic encryption is the fact that only the sources and the final destination of the data are capable of accessing the real data, thus preserving "end-to-end" confidentiality for the aggregation application scenario.

Additional security can be introduced by using probabilistic cryptosystems, such as the Pailler cryptosystem. In this case, given two encrypted values, it is not possible to decide whether they conceal the same value or not. This is especially useful to prevent eavesdroppers determining the content of a secure communication just by observing the encrypted packets.

5.2.5 Authorization Mechanisms for Secure IoT Services

Authorization mechanisms should be considered when deploying IoT services, in order to tackle the concerns that deployment of smart objects and services relying on them might raise in the minds of the public. In particular, authorization mechanisms must address the following questions:

- Which users can access some given data?
- How should the information be presented to a given user?
- Which operations is a user allowed to perform?

Role-based access control (RBAC) and attribute-based access control (ABAC) are the most common approaches to restricting system access to authorized users. RBAC maps permissions to roles that a user has been assigned. On the other hand, ABAC maps permissions to attributes of the user. However, authorization mechanisms strongly depend on an authentication step that must have been previously taken, in order to identify users. As an example, a complex, context-aware access control system designed for a medical sensor networks scenario, which has critical privacy and confidentiality issues, has been described by Garcia-Morchon and Wehrl [170].

Popular Internet-based services, such as social networks, have already faced the need to solve privacy-related problems when dealing with personal and protected data that might be accessed by third-parties; IoT applications are going to be facing the same issues. Since IoT services are expected to be offered in a RESTful paradigm, like those cited above, it may be helpful to borrow ideas from the experience that has been already created with Internet REST services.

The OAuth (Open Authorization) protocol was defined to solve the problem of allowing authorized third parties to access personal user data [171]. The OAuth protocol defines the following three roles.

- *Resource owner*: an entity capable of granting access to a protected resource, such as an end-user.
- *Resource server* (a service provider, SP): a server hosting user-related information.
- *Client* (a service sonsumer, SC): a third-party wanting to access personal user data to reach its goals.

An additional role is an authorization server (AS), which issues access tokens to the client after obtaining authorization from the resource owner.

In a general scenario, a SC that has been previously authorized by a user, can access the data that the user has made visible to the SC. This can be achieved by letting the SC retrieve the data from the SP on the user's behalf. In order to do so, one possible approach could be to force the user to give out personal authentication credentials to the SC. This approach has many drawbacks:

- the SC is going to appear to the SP just like the actual user, thus having unlimited access to the user's personal data;
- the user cannot define different restrictions for different SCs;
- the user cannot revoke the grant to a SC, unless it changes its credentials.

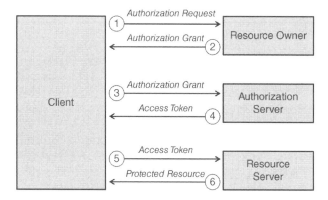

Figure 5.5 Interaction between the four roles of the OAuth protocol flow.

It is thus necessary to provide a mechanism that can separate the different roles. This can be done by granting specific credentials to the SC, which they can exhibit to the SP. These contain information about the SC's identity and the user's identity, so that the SP can serve its requests according to the access policies that the user has defined for the SC. The OAuth protocol defines the mechanisms that are needed to grant, use, and verify these credentials, which are called "access tokens".

The OAuth protocol defines the following flow of interaction between the four roles introduced above, as illustrated in Figure 5.5.

1) The client requests authorization from the resource owner: the authorization request can be made directly to the resource owner or, preferably, indirectly via the AS as an intermediary.
2) The client receives an authorization grant, which is a credential representing the resource owner's authorization.
3) The client requests an access token by authenticating with the AS and presenting the authorization grant.
4) The authorization server authenticates the client, validates the authorization grant, and, if the grant is valid, issues an access token.
5) The client requests the protected resource from the resource server and authenticates by presenting the access token.
6) The SP validates the access token and, if valid, serves the request.

The OAuth authorization framework (currently in version 2.0 [172]) enables a third-party application to obtain limited access to an HTTP service, either on behalf of a resource owner by orchestrating an approval interaction between the resource owner and the HTTP service, or by allowing the third-party application to obtain access on its own behalf.

For IoT scenarios, when using CoAP as an application-layer protocol instead of HTTP, a modified version of OAuth should be defined in order to ensure an authorization layer for restricting access to smart-object services. Since OAuth is an open protocol to allow secure authorization in a simple and standard method from web, mobile, and desktop applications, it has to be adapted in order to suit IoT application scenarios and to be compatible with constrained environments. For instance, header compression should be used to minimize the amount of information sent along.

HTTP/CoAP proxies should be able to perform OAuth proxying as well, in order to allow interoperability between conventional and constrained OAuth clients and servers. This raises particular challenges since the OAuth specification recommends the usage of HTTPS as a means to avoid man-in-the-middle attacks, thus preventing the access tokens from being stolen and used by malicious nodes. This means that CoAPs should be used in order to comply with the specification. The HTTP/CoAP proxy should then be able to perform a TLS-to-DTLS mapping in order to ensure end-to-end security. However, the use of HTTPS (and CoAPs inherently) can be avoided: it is possible to use OAuth over an insecure communication channel by adopting HMAC-SHA1 and RSA-SHA1 digital signature schemes.

5.3   Privacy Issues in the IoT

5.3.1   The Role of Authorization

The evolution of online services, such as those enabled by social networks, has had an important impact on the amount of data and personal information disseminated on the Internet. Furthermore, it has prompted the development of applications that use this information to offer new services, such as data aggregators. The information owned by online services is made available to third-party applications in the form of public application programming interfaces (APIs), typically using HTTP [2] as communication protocol and relying on the REST architectural style. The possibility that someone else besides the entity that generates the information and the service that is hosting it can access this information has raised concerns about the privacy of personal information, since the trust relationship is no longer pairwise, but possibly involves other parties, unknown at the time of service subscription.

Open Authorization (OAuth) is an open protocol to allow secure authorization from third-party applications in a simple and standardized way [173], as previously described. The OAuth protocol provides an authorization layer for HTTP-based service APIs, typically on top of a secure transport layer, such as HTTP-over-TLS (i.e., HTTPS) [174]. OAuth defines three main roles in the above scenario:

- the user (U) is the entity that generates information;
- the service provider (SP) hosts the information generated by the users and makes it available through APIs;
- the service consumer (SC), also referred to as the "client application", accesses the information stored by the SP for its own utilization.

In order to comply with the security and privacy requirements, U must issue an explicit agreement that some client application can access information on their behalf. This is achieved by granting the client an access token, containing U's and SC's identities, which must be exhibited in every request as an authorization proof. The OAuth 2.0 protocol is the evolution of the original OAuth protocol and aims at improving client development simplicity by defining scenarios for authorizing web, mobile, and desktop applications [175]. While connecting to existing online services is a simple task for client application developers, implementing an OAuth-based authorization mechanism on the SP side is a more complicated, time-consuming, and, potentially, computationally intensive task. Moreover, it involves the registration of users and client applications, and the permissions that users grant to SC applications, and their integration with authentication services.

International organizations, such as the IETF and the IPSO Alliance [176], and several research projects, such as the FP7 EU project CAL*I*PSO (Connect All IP-based Smart Objects!) [177], promote the use of IP as the standard for interoperability between smart objects.

The protocol stack run by smart objects tries to match classical Internet hosts in order to make it feasible to create the so-called "extended Internet"; in other words, the aggregation of the Internet with the IoT.

The IETF CoRE Working Group has defined the Constrained Application Protocol (CoAP; see Section 2.2.5.1) [7], a generic web protocol for RESTful constrained environments, targeted to M2M applications, which maps to HTTP for integration with the existing web.

Security in IoT scenarios is crucial. It applies at different levels, ranging from technological to privacy and trust issues, especially in scenarios involving smart toys (used by children) or crowd/social behavior monitoring. This is related to the fact that smart objects might have to deal with personal or sensitive data, which, if intercepted by unauthorized parties, might create ethical and privacy problems.

While the use of the OAuth protocol has little impact, in terms of processing and scalability, on conventional Internet-based services, its adoption in the IoT has to deal with the limitations and challenges of constrained devices. The limited computational power of smart objects may not be sufficient to perform the cryptographic primitives required for message authentication, integrity checks, and digital signatures, use of which would have a negative impact on energy consumption. Moreover, if the access permissions for the services provided by the smart object reside on the smart object itself, it could be extremely hard, if not impossible, to dynamically update them once they have been deployed. An example is found in smart parking systems, such as Fastprk[1] by Worldsensing [178], where smart objects may be embedded directly in the asphalt.

In rapidly evolving IoT scenarios, security is an extremely important issue. The heterogeneous and dynamic nature of the IoT raises several questions related to security and privacy, which must be addressed properly by taking into account the specific characteristics of smart objects and the environments they operate in. Classical security algorithms and protocols, used by traditional Internet hosts, cannot simply be adopted by smart objects, due to their processing and communication constraints. An extensive overview of state-of-the-art security mechanisms in the IoT (including symmetric/asymmetric cryptographic algorithms, hashing functions, security protocols at network/transport/application layers), aiming at providing features such as confidentiality, integrity, and authentication, can be found in the literature [179].

An architecture for solving the problem of securing IoT cyberentities (including smart objects, traditional hosts, and mobile devices),

---

1 http://www.fastprk.com/.

called U2IoT, has been proposed [180], with the goal of addressing the issues of expanding domains, dynamic activity cycles, and heterogeneous interactions. U2IoT takes into account security in interactions that occur in three different phases: preactive, active, and postactive. In particular, the active phase provides authentication and access control functionalities. Authorization is therefore being considered a major issue, since it is becoming increasingly evident that access to resources in a global-scale network, such as the IoT, must be controlled and restricted in order to avoid severe security breaches in deployed applications.

Several other studies have addressed very specific issues for the IoT. A lightweight multicast authentication scheme for small-scale IoT applications has been proposed [181]. Lai *et al.* take into account user mobility (i.e., roaming) and propose CPAL, an authentication mechanism designed to provide a "linking function" that can be used to enable authorized parties to link user access information, while preserving user anonymity and privacy [182]. The secure integration of WSNs into the IoT is discusses by Li and Xiong [183], who propose a security scheme that allows secure communication with Internet hosts by providing end-to-end confidentiality, integrity, and authentication, based on a public-key infrastructure. The proposed scheme also introduces a two-step (offline/online) signcryption mechanism, in order to minimize processing time.

Several authentication mechanisms have been defined for other issues, such as network access, that are also relevant for IoT scenarios. The Protocol for Carrying Authentication for Network Access (PANA) [184] is an IETF standard defining a network-layer transport for network access authentication. This is typically provided by the Extensible Authentication Protocol (EAP) [185]. PANA carries EAP, which can carry various authentication methods. OpenPANA [186] is an open-source implementation of PANA.

The problem of service authorization has been extensively treated in the literature. Several studies have focused on how to implement different access control strategies.

- Discretionary access control (DAC) restricts access to objects based on the identity of subjects and/or the groups to which they belong. The controls are discretionary in the sense that a subject with certain access permissions can transfer that permission on to any other subject [187].

- Role-based access control (RBAC) relies on a policy that restricts access to resources to those entities that have been assigned a specific role [188–190]: RBAC requires that the roles are defined and assigned to users, and that access permissions are set for resources.
- Attribute-based access control (ABAC) restricts resource access to those entities that have one or more specific attributes (e.g., age, geographic location, etc.) [191].

RBAC and ABAC are the most common approaches to restricting system access to authorized users. RBAC maps permissions to roles that a user has been assigned. On the other hand, ABAC maps permissions to attributes of the user. Typically, authorization mechanisms strongly depend on an authentication step that must have been previously taken in order to identify users so that either their roles or their attributes can be verified and matched against the policies set for resource access.

Schiffman *et al.* have presented a mechanism for fine-grained sub-delegation of access permissions for consumers of web applications, called DAuth. [192], Applying access control mechanisms in constrained scenarios, such as wireless sensor networks, is a challenging task.

IETF Authentication and Authorization for Constrained Environments (ACE) Working Group has also proposed the *Delegated CoAP Authentication and Authorization Framework* (DCAF) [193]. The DCAF architecture introduces authorization servers, which are used to perform authentication and authorization. Smart objects are prevented from having to store a large amounts of information by delegating the task to an external entity. While this solution is very similar to the one that we are about to describe, it focuses mainly on constrained environments, while the proposed one is intended to be generic and transparently integrated into both IoT and Internet scenarios.

In this section, we present a novel architecture, targeted at IoT scenarios, for an external authorization service based on OAuth, and called IoT-OAS. The delegation of the authorization functionalities to an external service, which may be invoked by any subscribed host or thing, affects:

- the time required to build new OAuth-protected online services, thus letting developers focus on service logic rather than on security and authorization issues;

- the simplicity of the smart object, which does not need to implement any authorization logic but must only securely invoke the authorization service in order to decide whether to serve an incoming request or not;
- the possibility to dynamically and remotely configure the access control policies that the SP is willing to enforce, especially in those scenarios where it is hard to intervene directly on the smart object.

Although much work has been done to define and integrate authorization mechanisms in several scenarios, the current proposal, unlike others, focuses on the definition of a generic authorization service that can be integrated into both Internet and IoT scenarios. In particular, the proposed mechanism explicitly takes into account the hybrid nature of the extended Internet that will be deployed in coming years. Moreover, the proposed architecture minimizes the effort required by service developers to secure their services by providing a standard, configurable, and highly interoperable authorization framework.

5.3.2 IoT-OAS: Delegation-based Authorization for the Internet of Things

5.3.2.1 Architecture

IoT-OAS can be invoked by any subscribed host or smart object. It can be thought of as a remotely triggered switch that filters incoming requests and decides whether to serve them or not. The design goal of the IoT-OAS architecture is to relieve smart objects of the burden of handling a large amount of authorization-related information and processing all incoming requests, even if unauthorized. By outsourcing these functionalities, smart objects can keep their application logic as simple as possible, thus meeting the requirements for keeping the memory footprint as low as possible, which is extremely important for constrained devices. From a broader perspective, entire large-scale IoT deployments can greatly benefit from the presence of IoT-OAS in terms of configurability: a single constrained node (or a group of constrained nodes as a whole) can have their access policies updated remotely and dynamically, without requiring any direct intervention, which is especially convenient for smart objects placed in hard-to-reach and/or unattended locations. OAuth allows third-party applications to get access to user-related information hosted on an online service. All issued requests must certify that the SC application has been granted permission by the user to access its personal information on its behalf, namely by adding an access token, which relates the user's identity and the client application. For ease of presentation, the acronyms used in this section are summarized in Table 5.6.

Besides the three roles introduced in Section 5.3.1 (U, SP, and SC), OAuth adds an additional role: the authentication service (AS), which is invoked by the SP to verify the identity of a user in order to grant access tokens. The standard OAuth operation flow is shown in

Table 5.6 Acronyms used.

U	User or resource owner
SP	Service provider, which hosts users' resources
SC	Service consumer, which accesses users' data stored by the SP
AS	Authentication service, which is used by the SP to verify the identity of a user
RT	Request token, a temporary ticket used by the SC to ask U to authorize access to its resources
AT	Access token, used by the SC to perform authenticated requests
IoT-OAS	Delegated external authorization service, which can be invoked by smart objects to perform authorization checks to access protected resources

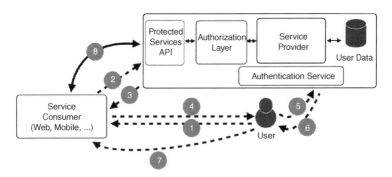

Figure 5.6 Standard OAuth roles and operation flow. The numbers indicate the sequence.

Figure 5.6. The procedure through which a SC can get a valid access token is the following:

1) U is willing to use the SC, either from a webpage, a mobile app, or a desktop application.
2) SC needs to access U's personal information hosted on the SP; SC asks the SP for an RT carrying SC's identity, which will be later exchanged for an AT.
3) SP verifies SC's identity and returns a RT.
4) SC redirects U to the SP's authentication service with the RT
5) U contacts the SP's AS presenting the RT and is asked to authenticate in order to prove its consent to grant access permissions to the SC.
6) The RT is exchanged for an AT, which relates U and SC.
7) The SC receives the AT through a redirection to a callback URL (i.e., authentication callback).
8) The SC can issue requests to SP including the AT, for services that require U's permission (protected APIs).

The design goal of the IoT-OAS architecture is to enable SPs, either based on HTTP or CoAP, to easily integrate an authorization layer without requiring any implementation overhead, other than invoking an external service. Delegating the authorization logic to an external service requires a strong trust relationship between the SP and the IoT-OAS. Figure 5.7 shows the operation flows for the AT grant procedure and the SC-to-SP interaction in the IoT-OAS architecture. A detailed description of these operation flows is presented in the remainder of this section.

5.3.2.2 Granting Access Tokens

The operation flow to grant an AT to a SC is shown in Figure 5.7a. The procedure resulting in the grant of an AT to a SC is similar to that of the standard OAuth operation flow, yet it has the following important differences:

1) As in the standard operation flow, the procedure is initiated by U.
2) The SC regularly contacts an SP to receive an RT.
3) The SP, which does not implement any OAuth logic, contacts the IoT-OAS asking for an RT for the SC by performing a generate_request_token RPC request.
4) The IoT-OAS verifies the identity of the SC and issues an RT, which is returned to the SP.

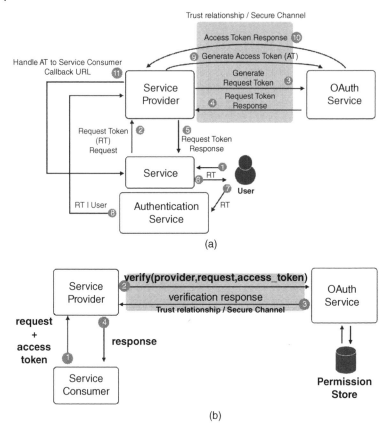

Figure 5.7 IoT-OAS main procedures: (a) AT grant procedure; (b) SP integration with IoT-OAS for request authorization. The numbers indicate the sequence.

5) The SP hands the RT back to the SC.

6) The SC redirects U to the AS with the received RT.

7) U contacts the SP's AS presenting the RT and authenticates it in order to prove consent to grant access permissions to the SC.

8) The AS notifies the SP that the authentication is successful and presents the RT with U's identity.

9) The SP asks the IoT-OAS to exchange the RT with an AT for U by issuing a generate_access_token RPC request.

10) The IoT-OAS generates the AT and returns it to the SP.

11) The SP hands the AT to the SC via an authentication callback.

The use of an external IoT-OAS is totally transparent to the SC, which has no knowledge of how the SP is implementing the OAuth protocol. This leads to full backward compatibility with standard OAuth client applications. On the SP side, all the OAuth logic is delegated to the IoT-OAS, with the only exception being the AS. However, it is not mandatory that the AS resides within the SP's realm, as it might interface with third-party authentication services, such as OpenID [194]. The only information the SP must hold is the reference to users' identities in order to make it possible to set up access permission policies on a per-user basis.

5.3.2.3 Authorizing Requests

The interaction between SP and IoT-OAS when serving incoming requests is shown in Figure 5.7b. Since the presence of the IoT-OAS is totally transparent to the SC, the communication between the SC and the SP is a regular OAuth communication. The difference is, again, on the side of the SP, which needs to contact the IoT-OAS to verify that the incoming requests received from the SC are authorized in order to decide whether to serve them or not.

The operation flow is as follows:

1) The SC requests U's information from the SP using the AT received after U's authentication (as in standard OAuth consumer-to-provider communication).
2) The SP, which does not implement any OAuth logic, refers to the IoT-OAS to verify if the incoming request is authorized (in order to do so, the SP issues a verify RPC request).
3) The IoT-OAS verifies the SC's request and informs the SP about the SC's authorization for the request by performing a lookup in the permission store.
4) The SP serves the SC's request according to the IoT-OAS's response.

5.3.2.4 SP-to-IoT-OAS Communication: Protocol Details

The SP interacts with the IoT-OAS with a simple communication protocol. The protocol comprises three remote procedure calls (RPCs), which are detailed below. It is important to note that delegating the authorization decision to an external service requires an extremely high trust level between the SP and the IoT-OAS. Moreover, all communications between them must be secured and mutually authenticated, so that the SP security level is at least as high as if the authorization service were implemented internally. To ensure that an appropriate security level is met, communications between the SP and the IoT-OAS must occur with a secure transport such as HTTP-over-TLS (HTTPS), CoAP-over-DTLS (CoAPs) [195, 196], or HTTP/CoAP over a secure host-to-host channel setup with IPSec [196]. Mutual authentication ensures that the IoT-OAS is verified and the requests come from a verified SP, whose identity is, therefore, implicit. The three RPCs of the SP-to-IoT-OAS communication protocol are as follows:

1) generate_request_token(): this RPC is called by the SP to request the IoT-OAS to generate a request token for the given SC, to be later exchanged for an AT.
2) generate_access_token(request_token,user_id): this RPC is called by the RPC to request the IoT-OAS to exchange the given RT for a new AT related to the given user.
3) verify(request,access_token): this RPC is called by the SP to request the IoT-OAS to verify if the given SC request is authorized with the provided AT by performing a lookup into the permission store.

5.3.2.5 Configuration

IoT-OAS provides high customization to SPs by offering per-user and per-service access control. The SPs can remotely configure and manage the permissions that SCs are granted, and these can be created, updated, revoked, and/or duplicated dynamically at any time. The IoT-OAS thus offers a dedicated SP access control configuration, known as the permission store. The permission store is a collection of the relations between SCs, users, and SP services and can be seen as a lookup table.

The configuration of the permission store can be through web interfaces or API calls provided by IoT-OAS. The possibility to dynamically manage the permissions, rather than having them co-located with the SP, is an extremely valuable feature, especially if SPs are smart objects with the need to be deployed in hard-to-access locations and that may be automatically and/or remotely reconfigured.

5.3.3 IoT-OAS Application Scenarios

In this section, we present four significant IoT application scenarios to illustrate the functionaly of the proposed IoT-OAS service architecture. We consider an external client (based on HTTP or CoAP according to the context) that wants to access a remote service provided by a network broker (NB), which is a border network element that exposes services on behalf of constrained nodes residing in the internal network. Alternatively, the service may be exposed directly by a generic smart object S directly available in the network behind a network gateway. To clarify the following description we assume that the OAuth credentials owned by the external client have been obtained through a prior configuration phase based on the IoT-OAS service described in Section 5.3.2 and that service discovery is performed through an application-specific procedure, which is not directly related to the approach presented in this work.

5.3.3.1 Network Broker Communication

In the first scenario, illustrated in Figure 5.8a, the client C (acting as an SC) discovers $Service_A$ provided by the NB. In order to serve external requests, the NB can retrieve information from different smart objects in its network. In this case, in order to simplify the context, we assume that the NB needs information only from the smart object S_2. The client, through a secure channel based on HTTPS or CoAPs, sends a request R for $Service_A$ including its OAuth credentials, denoted as $OAuth(C)$. The client's OAuth information is used by the service provider NB to validate the request, to verify the identity of C, and to confirm that C has the right privileges to access the requested service. Since NB could be implemented using an embedded device or, more generally, using a device with limited computational and storage capabilities (which, with high probability, does/should not implement a complex logic such as OAuth), it delegates the verification of the incoming request to the IoT-OAS. Once the NB receives R from C, it sends a verification request to the IoT-OAS (always through a secure channel based on HTTPS or CoAPs, according to its implementation or capabilities) with the original incoming request and its credentials. The IoT-OAS, after verifying the validity of the submitted request (according to NB's configuration) and C's identity, replies, communicating if R is authorized. If the feedback is positive and C is allowed to access the requested service, the NB internally contacts the smart object S_2 to retrieve the required information and sends back the response to C. If the response received from the IoT-OAS is negative, C is not granted access, and the NB responds immediately to C without any kind of interaction with the smart object.

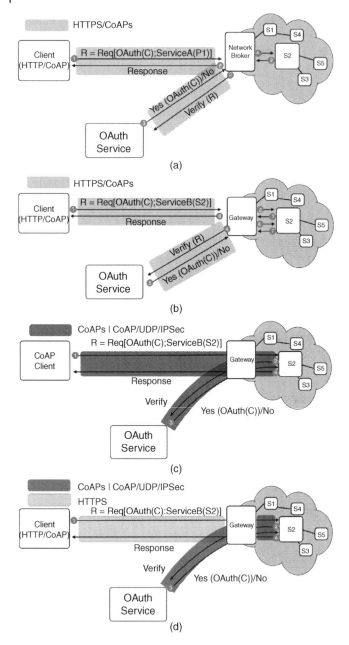

Figure 5.8 Application scenarios: (a) client-to-network broker communication; (b) gateway-enabled communication; (c) end-to-end CoAP communication between the external client and the smart object; (d) hybrid gateway-based communication.

5.3.3.2 Gateway-based Communication

In the second scenario, illustrated in Figure 5.8b, an external client C, based on HTTPS or CoAPs communications, is interested in accessing a service directly provided by the smart object S_2, which does not run HTTP or CoAP (due to computational or implementation constraints) and is behind a gateway G. Gateway G has the role of translating the incoming requests from external networks to available smart objects inside its own network. In this scenario, C sends a request R (including the client's OAuth credentials) to G for $Service_B$ provided by S_2. R is translated by G and forwarded to S_2, which, in order to validate the new request and the requestor's identity, sends through G a verification request to the IoT-OAS using a secure communication protocol such as HTTPS or CoAPs. The verification message and the response (positive or negative) generated by S_2 are managed and translated by G to allow communication between IoT-OAS, the smart object, and the client.

5.3.3.3 End-to-End CoAP Communication

Figure 5.8c shows a different scenario, where a smart object S_2 (i.e., reachable at an IPv6 address) in a sensor network directly provides a remote CoAP service $Service_B$. Since all the involved entities can use the same protocol, in this case the network gateway acts only as a router without the need to translate incoming and outgoing messages between the external world and the sensor network. The CoAP client CC sends securely and directly to the smart object a request R containing its OAuth credentials and the reference for $Service_B$ provided by S_2. Since the smart object is usually a sensor or an embedded device with limited computational and storage capabilities (and therefore, as previously described, does not implement a complex logic like OAuth), it delegates the verification of the incoming request to the OAuth service. S_2 sends a verification request to IoT-OAS over CoAPs to check R. The IoT-OAS validates the request based on CC's credentials and the type of requested service; it then informs the smart object S_2 about whether R can be served or not. S_2, according to the response of IoT-OAS, replies to the requesting client with the service outcome or, if the CC is not allowed to access $Service_B$, with an error message.

5.3.3.4 Hybrid Gateway-based Communication

The last scenario, shown in Figure 5.8d, is characterized by a hybrid approach in which the external client uses an application protocol (such as HTTP) that is different from that used by smart objects (CoAP). Similar to the case in Section 5.3.3.2, the gateway manages the communication between the external world and its network: in this case, it just translates incoming requests from HTTP to CoAP for S_2. Once a new request (with OAuth credentials and service reference) arrives at the smart object, it uses IoT-OAS securely to verify the validity of R, through CoAPs. The response (positive or negative, according to the IoT-OAS feedback) is translated by the gateway from CoAP to HTTP and forwarded to the client through a secure channel.

6

Cloud and Fog Computing for the IoT

6.1 Cloud Computing

Cloud Computing is an increasing trend, involving the move to external deployment of IT resources, which are then offered as services [197]. Cloud Computing enables convenient and on-demand network access to a shared pool of configurable computing resources (e.g., networks, servers, storage elements, applications, and services). These can be rapidly provisioned and released with minimal management effort or service provider interaction [198]. At the hardware level, a number of physical devices, including processors, hard drives, and network devices, fulfill processing and storage needs. Above this, a combination of a software layer, a virtualization layer, and a management layer, allows effective management of servers.

The available service models are as follows:

- *Infrastructure as a Service* (IaaS): provides processing, storage, networks, and other computing resources, allowing the consumer to deploy and run arbitrary software, including OSs and applications. The consumer has control over the OSs, storage, deployed applications and, possibly, limited control of select networking components.
- *Platform as a Service* (PaaS): provides the capability to deploy infrastructure and consumer-created or acquired applications. The consumer has no control over the underlying infrastructure (e.g., network, servers, OSs, or storage) but only manages deployed applications.

Internet of Things: Architectures, Protocols and Standards, First Edition.
Simone Cirani, Gianluigi Ferrari, Marco Picone, and Luca Veltri.
© 2019 John Wiley & Sons Ltd. Published 2019 by John Wiley & Sons Ltd.

- *Software as a Service* (SaaS): provides the capability to use the provider's applications, running on the cloud infrastructure. These application are accessed from client devices through suitable client interfaces. The consumer does not manage or control the underlying cloud infrastructure or individual application capabilities, with the possible exception of limited user-specific application configuration settings.

Cloud computing is generally complementary to the IoT scenario, as it acts

- as a collector of real-time sensed data
- as provider of services built on the basis of collected information.

The main need is to be extremely scalable, thus allowing support for large-scale IoT applications. There are several open-source frameworks and technologies which can be used for cloud IoT systems, such as OpenStack (created by Rackspace and NASA in 2010) and OpenNebula [199]. The former is an open cloud OS that controls large pools of computing, storage, and networking resources and can be seen as a framework with a vendor-driven model; the second is an open-source project aimed at delivering a simple, feature-rich, and flexible solution to build and manage enterprise clouds and virtualized data centers.

6.2 Big Data Processing Pattern

From a business perspective, managing and gaining insights from data is a challenge and a key in gaining competitive advantage. Analytical solutions that mine structured and unstructured data are important, as they can help companies to gain information, not only from their privately acquired data, but also from the large amounts of data publicly available on the web, social networks, and blogs. Big Data opens up a wide range of possibilities for organizations to understand the needs of their customers, predict their demands, and optimize valuable resources.

Big Data is different to and more powerful than traditional analytics tools used by companies [200]: it can find patterns and glean intelligence from data, translating them into business advantage. However,

Big Data is powered by what is often referred as a "multi-V" model, in which V stands for:

- *variety*: to represent the data types;
- *velocity*: to represent the rate at which the data is produced and processed and stored according with further analysis;
- *volume*: to define the amount of data;
- *veracity*: refers to how much the data can be trusted given the reliability of its sources.

Big Data architectures generally use traditional processing patterns with a pipeline approach [201]. These architectures are typically based on a processing perspective: the data flow goes downstream from input to output, to perform specific tasks or reach the target goal. Typically, data are sequentially handled with tightly coupled pre-defined processing sub-units (static data routing). The paradigm can be described as "process-oriented": a central coordination point manages the execution of subunits in a certain order and each sub-unit provides a specific processing output, which is used only within the scope of its own process, without the possibility of being shared among different processes. This approach represents a major deviation from traditional service-oriented architectures, where the sub-units are external web services invoked by a coordinator process rather than internal services [202]. Big Data applications generally interact with cloud computing architectures, which can handle resources and provide services to consumers. Assunção *et al.* have provided a survey of approaches, environments, and technologies in key areas for big data analytics, investigating how they can contribute to build analytics solutions for clouds [203]. Gaps in knowledge and recommendations for the research community on future directions for cloud-supported big data computing are also described.

6.3 Big Stream

Billions of smart objects are expected to be deployed in urban, home, industrial, and rural scenarios, in order to collect information, which might then be used to build new applications. Shared and interoperable communication mechanisms and protocols are currently being defined and standardized, allowing heterogeneous nodes to efficiently communicate with each other and with existing Internet actors. An IP-based IoT will be able to extend and interoperate seamlessly with the existing Internet. Standardization institutions, such as the IETF [204], and several research projects [177] are in the process of defining mechanisms to bring IP to smart objects, a result of the need to adapt higher-layer protocols to constrained environments. However, not all objects will support IP, as there will always be tiny devices that are organized in closed/proprietary networks and rely on very simple and application-specific communication protocols. These networks will eventually connect to the Internet through a gateway/border router.

Sensed data are typically collected and sent by uplink, namely from IoT networks, where smart objects are deployed, to collection environments (server or cloud), possibly through an intermediate local network element, which may perform some processing tasks, such as data aggregation and protocol translation. Processing and storing data at the edge of networks (e.g., on set-top boxes or access points) is the basis for the evolution of fog computing [205] in IoT scenarios. Fog Computing is a novel paradigm that aims at extending cloud computing and services to the edge of the network, leveraging its proximity to end users, its dense geographical distribution, and its support for mobility. The wide geographical distribution makes the fog computing paradigm particularly suited to real-time big data analytics. Densely distributed data collection points allow a fourth axis – low-latency – to be added to typical big data dimensions (volume, variety, and velocity). Figure 6.1 shows the hierarchy of layers involved in data collection, processing, and distribution in IoT scenarios.

With billions of nodes capable of gathering data and generating information, the availability of efficient and scalable mechanisms for collecting, processing, and storing data is crucial. Big Data techniques, which were developed over the last few years (and became popular due to the evolution of online and social/crowd services), address the need to process extremely large amounts of heterogeneous data for multiple purposes. These techniques have been designed mainly to deal with huge volumes (focusing on the amount of data itself), rather than providing real-time processing and dispatching. Cloud Computing has been direct application for big data analysis due to its scalability, robustness, and cost-effectiveness. The number of data sources, on one side, and the subsequent frequency of incoming data, on the other side, create a new need for cloud architectures to handle massive flows of information, thus producing a shift from the big data

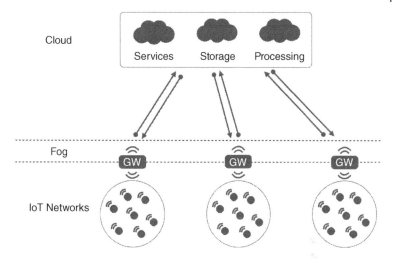

Figure 6.1 The hierarchy of layers involved in IoT scenarios: the fog works as an extension of the cloud to the network edge, where it can support data collection, processing, and distribution.

paradigm to the big stream paradigm. In addition, the processing and storage functions implemented by remote cloud-based collectors are the enablers for their core businesses: providing services based on the collected/processed data to external consumers.

Several relevant IoT scenarios, (such as industrial automation, transportation, networks of sensors and actuators), require real-time/ predictable latency and could even change their requirements (e.g., in terms of data sources) dynamically and abruptly. Big-stream-oriented systems will be able to react effectively to changes and to provide smart behavior when allocating resources, thus implementing scalable and cost-effective cloud services. Dynamism and real-time requirements are another reason why big data approaches, with their intrinsic inertia (big data typically uses batch processing), are not suitable for many IoT scenarios.

The main differences between the Big Data and Big Stream paradigms are the nature of the data sources and the real-time/latency requirements of the consumers. The big stream paradigm allows real-time and ad-hoc processing to be performed, to link incoming streams of data to consumers. It offers a high degree of scalability, fine-grained and dynamic configuration, and management of heterogeneous data formats. In brief, while both Big Data and Big Stream

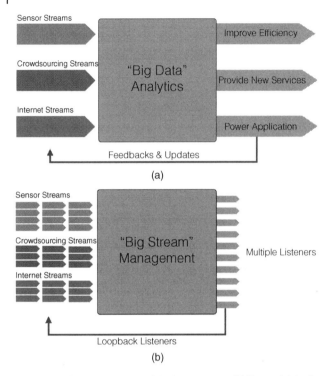

Figure 6.2 (a) Data sources in big data systems. (b) The multiple data sources and listener management in big stream systems.

systems deal with massive amounts of data, the former focuses on the analysis of data, while the latter focuses on the management of flows of data, as shown in Figure 6.2.

The difference is in the meaning of the term "big", which refers to the volume of data for big data and to the global data generation rate of the data sources for big stream. This difference is also relevant for the data that are considered relevant to consumer applications. For instance, while for big data applications it is important to keep all sensed data in order to be able to perform any required computation, in big stream applications, data aggregation or pruning may be performed in order to minimize the latency in conveying the results of computations to consumers; there is no need for persistence. Note that, as a generalization, big data applications might be consumers of big stream data flows.

For these reasons, we present an architecture targeting cloud-based applications with real-time constraints – big stream applications – for IoT scenarios. It relies on the concepts of the data listener and the data-oriented processing graph in order to implement a scalable, highly configurable, and dynamic chain of computations on incoming big streams and to dispatch data using a push-based approach, thus providing the shortest delay between the generation of information and its consumption.

6.3.1 Big-stream-oriented Architecture

As stated in Section 6.3, a major difference between big data and big stream approaches is the real-time/low-latency requirements of big stream consumers. The vast number of data sources in IoT applications has made cloud-service implementors mistakenly believe that re-using big data-driven architectures will be the right solution for all applications, and that there is no need to design new paradigms specific for IoT scenarios. IoT application scenarios are characterized by a huge number of data sources sending small amounts of information to a collector service, typically at a limited rate. Many services can be built upon these data: environmental monitoring, building automation, and smart cities applications are just a few examples. These applications typically have real-time or low-latency requirements if they are to provide efficient reactive/proactive behavior. This could be effectively implemented in an IP-based IoT, where smart objects can be directly addressed.

Applying a traditional big data approach for IoT application scenarios might bring higher or even unpredictable latency between data generation and its availability to a consumer, since this was not among the main objectives behind the design of big data systems. Figure 6.3 illustrates the time contributions introduced when data pushed by smart objects need to be processed, stored, and then polled by consumers. The total time required by any data to be delivered to a consumer can be expressed as $T = t_0 + t_1 + t_2$, where:

- t_0 is the time elapsed from the moment a data source sends information, through an available API, to the cloud service (1) and the service dispatches the data to an appropriate queue, where it can wait for an unpredictable time (2), in order to decouple data acquisition from processing.

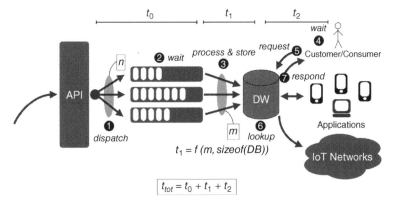

Figure 6.3 Traditional big data architecture for IoT and delay contributions from data generation to applications information delivery. Refer to text for details.

- t_1 is the time needed for data, extracted by the queue, to be pre-processed and stored in data warehouse (DW) (3); this time depends on the number of concurrent processes that need to be executed and get access the common DW and the current size of the DW.
- t_2 is the data consumption time, which depends on the remaining time that a polling consumer needs to wait before performing the next fetch (4), the time for a request to be sent to the cloud service (5), the time required for lookup in the DW and post-processing of the fetched data (6), and the time for the response to be delivered back to the consumer (7).

It can be seen that the architecture described is not optimized to minimize the latency – and, therefore, to feed (possibly a large number of) real-time applications – but, rather, to perform data collection and batch processing. Moreover, it is important to understand that data significant for Big Stream applications might be short-lived, since they are to be consumed immediately, while big data applications tend to collect and store massive amounts of data for an unpredictable time.

In this chapter, we outline a novel architecture explicitly designed for the management of big stream applications targeting IoT scenarios. The main design criteria are the minimization of the latency in data dispatching to consumers and the optimization of resource allocation. The main novelty in the proposed architecture is that the data flow is "consumer-oriented", rather than being based on the knowledge of collection points (repositories) where data can be retrieved.

The data being generated by a deployed smart object might be of interest for a consumer application, termed the "listener". A listener registers its interest in receiving updates (either in the form of raw or processed data) coming from a streaming endpoint (i.e., a cloud service). On the basis of application-specific needs, each listener defines a set of rules to specify what type of data should be selected and the associated filtering operations. For example, in a smart parking application, a mobile app might be interested in receiving content related only to specific events (e.g., parking sensors status updates, the positions of other cars, weather conditions, and so on) that occur in a given geographical area, in order to accomplish relevant tasks, such as finding a free, covered parking spot. The pseudocode that can be used to express the set of rules for the smart parking application is shown in Listing 6.1:

Listing 6.1 Smart parking pseudocode.

```
when
    $temperatureEvent = {
        @type:http://schema.org/Weather#temperature}
    $humidityEvent = {
        @type:http://schema.org/Weather#humidity}
    $carPositionEvent = {
        @type:http://schema.org/SmartCar#travelPosition}
    $parkingStatusEvent = {
        @type:http://schema.org/SmartParking#status
    }
    @filter: {
      location: {
      @type:"http://schema.org/GeoShape#polygon",
      coordinates: [ [
        [41.3983, 2.1729], [41.3986, 2.1729],
        [41.3986, 2.1734], [41.3983, 2.1734],
        [41.3983, 2.1729]
      ] ]
    }
then
    <application logic>
```

The rules specify:

- what kinds of events are of interest for the application;
- a geographical filter to apply in order to receive only events related to a specific area.

Besides the final listener (end-user), the cloud service might also act as a listener and process the same event data stream, but with different rules, in order to provide a new stream that can be consumed by other listeners. An example would be a real-time traffic information application, the pseudo-code for which is presented in the following listing:

Listing 6.2 Traffic information application pseudocode.

```
when
    $cityZone = {@type:http://schema.org/SmartCity#zone}
    $carPositionEvents = collect({
        @type: http://schema.org/SmartCar#travelPosition,
        @filter: {
                    location: cityZone.coordinates
        }
    }) over window:time(30s)
then
    emit {
        @type: http://schema.org/SmartCity#trafficDensity,
        city_zone: $cityZone,
        density: $carPositionEvents.size,
    }
```

The proposed big stream architecture guarantees that, as soon as data are available, they will be dispatched to the listener, which is thus no longer responsible for polling data, thus minimizing latency and possibly avoiding network traffic.

The information flow in a listener-based cloud architecture is shown in Figure 6.4. With the new paradigm, the total time required by any data to be delivered to a consumer can be expressed as:

$$T = t_0 + t_1 \tag{6.1}$$

where:

- t_0 is the time elapsed from the moment a data source sends information, through an available API, to the cloud service (1) and the service dispatches the data to an appropriate queue, where it can wait for an unpredictable time (2), in order to decouple the data acquisition from processing.
- t_1 is the time needed to process data extracted from the queue and be processed (according to the needs of the listener, say to perform format translation) and then deliver it to registered listeners.

It is clear that the reversed perspective introduced by listener-oriented communication is optimal in terms of minimization of the time that

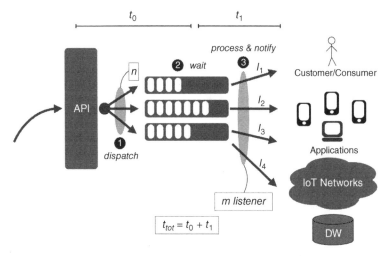

Figure 6.4 The proposed listener-based architecture for the IoT: delay contributions from data generation to consumer information delivery are explicitly indicated.

a listener must wait before it receives data of interest. In order to highlight the benefits brought by the big stream approach over a big data approach, consider an alerting application, which should notify one or more consumers of an event in the fastest possible time. The traditional big data approach would require an unnecessary pre-processing/storage/post-processing cycle to be executed before the event could be made available to consumers, who would be responsible to retrieve the data by polling. The listener-oriented approach, on the other hand, guarantees that only necessary processing will be performed before data are delivered directly to the listener, thus providing an effective real-time solution.

This general discussion shows that a consumer-oriented paradigm is better suited to real-time Big Stream applications than simply reusing existing big data architectures, which best suit applications that do not have critical real-time requirements.

6.3.2 Graph-based Processing

In order to overcome the limitations of the "process-oriented" approach described in Section 6.2, and fit the proposed Big Stream

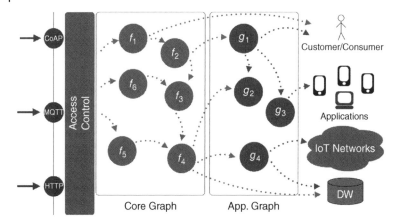

Figure 6.5 The proposed listener-based graph architecture: the nodes of the graph are listeners and the edges are the dynamic flow of information data streams.

architecture, we have envisioned and designed a new cloud graph-based architecture built on top of basic building blocks that are self-consistent and perform "atomic" processing on data, but that are not directly linked to a specific task. In such systems, the data flows are based on dynamic graph-routing rules determined only by the nature of the data itself and not by a centralized coordination unit. This new approach allows the platform to be "consumer-oriented" and to give optimal resource allocation. Without the need for a coordination process, the data streams can be dynamically routed in the network by following the edges of the graph. This allows the possibility of automatically switching off nodes when processing units are not required, or transparently replicating nodes if some processing entity is overwhelmed by a significant number of concurrent consumers.

Figure 6.5 illustrates the proposed directed-graph processing architecture and the concept of the listener. A listener is an entity (e.g., a processing unit in the graph or an external consumer) interested in the raw data stream or in the output provided by a different node in the graph. Each listener represents a node in the topology and the presence and combination of multiple listeners, across all processing units, defines the routing of data streams from producers to consumers. In this architectural approach:

- nodes are processing units (processes), performing some kind of computation on incoming data;
- edges represent flows of information, linking together various processing units, which are thus together able to implement some complex behavior;
- nodes of the graph are listeners for incoming data or outputs of other nodes of the graph.

The graph-based approach allows resource allocation to be optimized in terms of

- efficiency, by switching off processing units that have no listeners registered to them (enabling cost-effectiveness)
- scalability, by replicating those processing units which have a large number of registered listeners.

The combination of these two functionalities and the concept of the listener allow the platform and the overall system to adapt itself to dynamic and heterogeneous scenarios by properly routing data streams to the consumers, and to add new processing units and functionalities on demand.

In order to provide a set of commonly available functionalities, while allowing for dynamic extension of the capabilities of the system, the graph comprises several concentric levels:

- a core graph, with basic processing provided by the architecture (e.g., format translation, normalization, aggregation, data correlation, and other transformations);
- one or more application graphs, with listeners that require data coming from an inner graph level in order to perform custom processing on already processed data.

The complexity of processing is directly proportional to the number of levels crossed by the data. This means that data at an outer graph level must not be processed again at an inner level. From an architectural viewpoint, as shown in a scheme in Figure 6.6, nodes at inner graph levels cannot be listeners of nodes of outer graph levels. In other words, there can be no link from an outer graph node to an inner graph node, but only vice versa. Same-level graph nodes can be linked together if there is a need to do so. Figure 6.7 illustrates incoming and outgoing listener flows between core- and application-graph units. In particular, a processing unit of the core graph can be a listener only

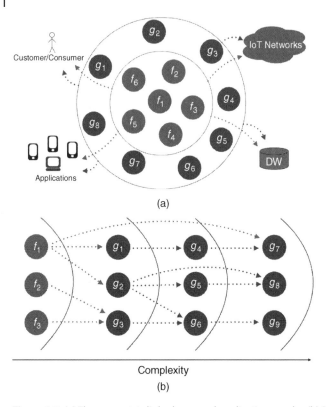

(a)

(b)

Complexity

Figure 6.6 (a) The concentric linked core and application graphs. (b) Basic processing nodes build the core graph: the outer nodes have increasing complexity.

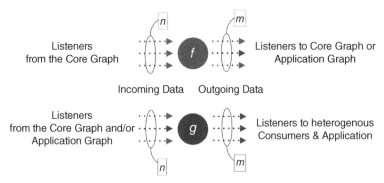

Figure 6.7 Allowed input and output flows for core-graph and application-graph nodes.

for other nodes of the same level (n incoming streams) and a source both for other core or application graph nodes (m outgoing streams). A node of an application graph can be, at the same time:

- a listener to n incoming flows from core and/or application graphs;
- a data source only for other (m) nodes of the application graph or heterogeneous external consumers.

6.3.3 Implementation

In this section, the functionalities and the details of the implementation of the proposed architecture using standard protocols and open-source components are presented. Three main modules form the system:

- acquisition and normalization of the incoming raw data;
- graph management;
- application register entity.

All modules and their relationships are shown in Figure 6.8. A detailed explanation is given in the following sections.

6.3.3.1 Acquisition Module

The acquisition module represents the entry point, for external IoT networks of smart objects, to the cloud architecture. Its purpose is to receive incoming raw data from heterogeneous sources, making them available to all subsequent functional blocks. As mentioned

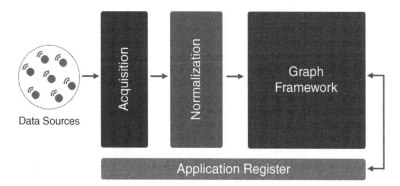

Figure 6.8 Components of the proposed graph-based cloud architecture and relations between each element.

before about IoT models, several application-layer protocols can be implemented by smart objects; adhering to this idea, the acquisition module has been modeled to include a set of different connectors, in order to be able to handle each protocol-specific incoming data stream. Among the most common IoT application-layer protocols, the current implementation of the acquisition module supports HTTP, CoAP and MQTT. In order to increase scalability and efficiency, in the module implementation an instance of NGINX [206] has been adopted as an HTTP acquisition server node. The server is reachable via the default HTTP port, working with a dedicated PHP page as processing module. The latter has been configured to forward incoming data to the inner queue server. We have chosen NGINX, instead of the prevailing and well-known open-source Apache HTTPD Server [207], because it uses an event-driven asynchronous architecture to improve scalability and, specifically, aims to guarantee high performance even in the presence of a critical number of requests. The CoAP acquisition interface was implemented using a Java process, based on a mjCoAP server instance [24], waiting for incoming raw messages, and connected to the RabbitMQ queue server,[1] passing it injected elements. Indeed, since the proposed architecture is big-stream-oriented, a suitable messaging paradigm is queue communication, so in the developed platform an instance of RabbitMQ queue broker was adopted. The MQTT acquisition node is built by implementing an Apache Software Foundation ActiveMQ server through a Java process that listens for incoming data over a specific input topic (mqtt.input). This solution was preferred to other solutions (e.g., the C-based server Mosquitto) because it provides a dedicated API that allows custom development of the component. The MQTT acquisition node is also connected to the architecture's queue server. In order to avoid potential bottlenecks and collision points, each acquisition protocol module has a dedicated exchange module and queue (managed by RabbitMQ), linked together with a protocol-related routing key, ensuring the efficient management of incoming streams and their availability to the subsequent nodes. In the described implementation, an exchange is a RabbitMQ component that acts as a router in the system and dispatches incoming messages to one or more output queues, following dynamic routing rules.

---

1 `www.rabbitmq.com`.

6.3.3.2   Normalization Module

Since incoming raw data are generally application- and theme-dependent, a normalization module has been incorporated in order to normalize all the collected information and generate a representation suitable for processing. The normalization procedure involves fundamental and atomic operations on data, such as:

- suppression of useless information (e.g., unnecessary headers or metadata);
- annotation with additional information;
- translation of payloads to a suitable format.

In order to handle the huge amount of incoming data efficiently, the normalization step is organized with protocol-specific queues and exchanges. As shown in the normalization section of Figure 6.9, the information flow originated by the acquisition module is handled as follows. All protocol-specific data streams are routed to a dedicated protocol-dependent exchange, which forwards them to a specific queue. A normalization process handles the input data currently available on that queue and performs all necessary normalization operations in order to obtain a stream of information units that can be processed by subsequent modules. The normalized stream is forwarded to an output exchange. The main advantage of using exchanges is that queues and normalization processes can be dynamically adapted to the current workload. For instance, normalization

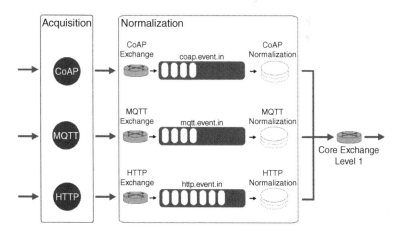

Figure 6.9 Detailed representation of acquisition and normalization blocks.

queues and processes can be easily replicated to avoid system congestion. Each normalization node has been implemented as a Java process, analyzing incoming raw data extracted from a queue identified through a protocol-like routing key (e.g., <protocol>.event.in), leaving unaltered the associated routing key, which identifies the originator smart object's protocol. The received data are fragmented and encapsulated in a JSON-formatted document, which provides an easy-to-manage format. At the end of the normalization chain, each processor node forwards its new output chunk to the next exchange, which represents the entry-point of the graph module, promoting data flows to the next layers of the proposed architecture.

6.3.3.3 Graph Framework

The graph framework comprises a number of different computational processes representing a single node in the topology; layers are linked together with frontier exchanges, forwarding data streams to their internal nodes.

Each graph node i of a specific layer n is a listener, waiting for an input data stream on a dedicated layer-n exchange-connected queue. If this node also acts as publisher, after performing its processing on the input data, it can deliver computation results to the its layer-n exchange. In order to forward streams, information generated by node i become available for layer n and layer $n + 1$ listeners that are interested in this kind of data, thanks to the binding between layer-n and layer($n + 1$) exchanges.

Incoming messages are stored in active queues, connected to each graph layer's exchange. Queues can be placed into the core graph layers, for basic computation, or into application graph layers, for enhanced computation. Layers are connected through one-way links to their successor exchange, using the binding rules allowed by the queue manager, ensuring proper propagation of data flows and avoiding loops. Each graph layer is composed by Java-based graph nodes dedicated to process data provided by the graph layer's exchange. Such nodes can either be core, if they are dedicated to simple and primitive data processing, or application, if they are oriented to a more complex and specific data management.

Messages, identified with a routing key, are first retrieved from the layer's exchange, are then processed, and finally are sent to the target exchange, with a new work-related routing key, as shown in Figure 6.10. If the outgoing routing key belongs to the graph layer

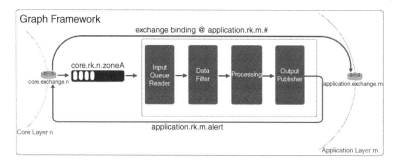

Figure 6.10 Interaction between core and application layers with binding rule.

as the incoming message, the data remain in the same exchange and become available for other local processes. If the outgoing routing key belongs to an outer graph layer, then data are forwarded to the corresponding exchange and finally forwarded, adhering to binding rules. Each graph node, upon becoming part of the system, can specify if it acts as a data publisher, capable of handling and forwarding data to its layer's exchange, or if it acts only as data consumer. Data flow continues until it reaches the last layer's exchange, which is responsible for managing notifications to external entities that are interested in the final processed data, such as data warehouses, browsers, smart entities, and other cloud graph processes.

6.3.3.4 Application Register Module

The application register module has the fundamental responsibilities:

- to manage the processing graph by maintaining all the information about the current statuses of all graph nodes in the system
- to route data across the graph.

In more detail, the application register module performs the following operations:

- attach new nodes or consumer applications interested in streams provided by the system;
- detach nodes of the graph that are no longer interested in streaming flows and eventually reattach them if required;
- deal with nodes that are publishers of new streams;
- maintain information regarding data topics, in order to correctly generate the routing keys and to compose data flows between nodes in different graph layers.

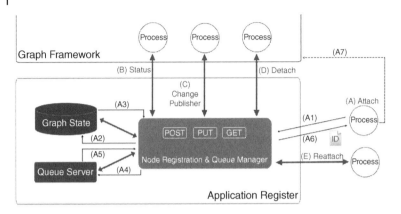

Figure 6.11 Detailed representation of the application register module, with possible actions that may be performed by graph nodes, highlighting ATTACH request steps needed to include an external node in the graph.

In order to accomplish all these functionalities, the application register module is made up of two main components, as shown in Figure 6.11.

The first module is the graph state database, which stores all the information about active graph nodes, such as their states, layers, and whether they are publishers. The second module is the node registration and queue manager (NRQM), which handles requests from graph nodes or external processes, and handles queue management and routing in the system. When a new process joins the graph as a listener, it sends an attach request to the application register module, specifying the kind of data that it is interested to. The NQRM module stores the information of the new process in the graph state database and creates a new dedicated input queue for the process, according to its preferences. Finally, the NRQM sends a reference of the queue to the process, which becomes a new listener of the graph and can read the incoming stream from the input queue.

After this registration phase, the node can perform new requests (e.g., publish, detach, and get status). The overall architecture is managed by a Java process (the application register), which coordinates the interactions between graph nodes and external services, such as the RabbitMQ queue server and the MySQL database. It maintains and updates all information and parameters related to processing unit queues. As a first step, the application register starts up all the external connections, and then it activates each layer's exchange, binding them with their successors. Finally, it proceeds with the activation of a Jetty HTTP server, responsible for listening and handling all core and application node requests, as shown in Figure 6.11 using a RESTful HTTP paradigm. These requests are:

- (A) attach
- (B) status
- (C) change publishing policy
- (D) detach
- (E) re-attach.

Figure 6.12 shows all of the architecture modules described above, along with a detailed indication of the information flows.

6.3.4 Performance Evaluation

The implementation of the proposed graph framework for big stream management was carried out by deploying an Oracle VirtualBox VM, equipped with Linux Ubuntu 12.04 64-bit, 4 GB RAM, two CPUs and 10 GB HDD. The implemented architecture was evaluated through the definition of a real use case: a smart parking scenario. The data traces used for the evaluation of the proposed architecture were provided by WorldSensing from one of the company's deployments in a real-life scenario, used to control parking spots on streets. The traces are a subset of an entire deployment (more than 10,000 sensors) with information from 400 sensors over a three-month period, forming a dataset with more than 604,000 parking events. Each dataset item is represented by:

- sensor ID
- event sequence number relative to the specific sensor
- event timestamp
- parking spot status (free/busy).

No additional information about the parking zone was provided. Therefore, in order to create a realistic scenario, parking spot sensors were divided into seven groups, representing different parking zones of a city. This parking-spot–city-zone association was stored in an external database.

The parking dataset was used in the cloud infrastructure using a Java-based data generator, which simulates the IoT sensor network. The generator randomly selects an available protocol (HTTP, CoAP,

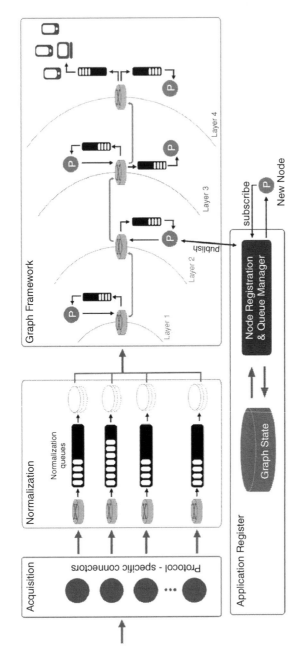

Figure 6.12 The complete graph cloud architecture, with reference to the data stream flows between all building blocks, from IoT data sources to final consumers.

or MQTT) and periodically sends streams to the corresponding acquisition node interface. Once the data has been received by the acquisition layer, they are forwarded to the dedicated normalization exchange, where corresponding nodes enrich incoming data with platform-specific details. With reference to the selected scenario, the normalization stage adds parking zone details to the input data, retrieving the association from an external database. Once the normalization module has completed its processing, it sends the structured data to the graph framework, allowing further processing of the enriched data stream.

The graph framework considered in our experimental set-up comprises eight core layers and seven application layers, within which different node topologies are built and evaluated.

Processed data follow a path based on routing keys, until the final external listener is reached. Each application node is interested in detecting changes of parking-spot data related to specific parking zones. Upon a change of the status, the graph node generates a new aggregated descriptor, which is forwarded to the responsible layer's exchange, which has to notify the change event to external entities interested in the update (free → busy, busy → free). The rate of these events, coming from a real deployment in a European city, respects some rules imposed by the company, and for our purposes might seem low. Thus, in order to cause sufficient stress to the proposed big stream cloud system, the performance was evaluated by varying the data generation rate in a suitable range. In other words, we force a specific rate of incoming events, without taking into account the real parking spots timestamps gathered from the dataset.

The proposed architecture was evaluated using the testbed described in Section 6.3.3, by varying the incoming raw data from 1 msg/s to 100 msg/s. The evaluation involved assessing the performance of the acquisition stage and the computation stage. First, performance was evaluated by measuring the time difference between the instant at which data were sent from a data generator to the corresponding acquisition interface and the instant at which the data were enriched by normalization nodes, thus becoming available for processing by the first core node. The results are shown in Figure 6.13. The acquisition time slightly increased but it was around 15 ms at all rates considered.

The second performance evaluation measured the time between the instant at which enriched data became ready for processing activities

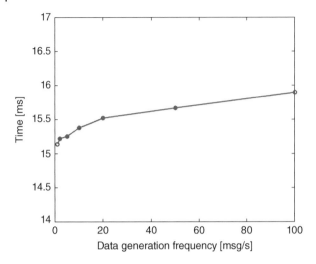

Figure 6.13 Average time related to the acquisition block.

and the instant at which the message reached the end of its graph framework routes, becoming available for external consumers/customers. In order to consider only the effective overhead introduced by the architecture, and without considering implementation-specific contributions, performance results were obtained by subtracting the processing time of all core and application nodes. Finally, these times were normalized over the number of computational nodes, in order to obtain the per-node overhead introduced by the architecture, independent of the specific routing and topology that were implemented. The results, shown in Figures 6.14 and 6.15, were calculated using the following expression:

$$T_{\text{processing}_{\text{freq}}} = \frac{T_{\text{out}} - T_{\text{in}} - \sum_{k=1}^{N} GP_k}{N} \tag{6.2}$$

where: T_{out} is the instant at which parking data reach the last application layer; T_{in} is the instant at which normalized data comes to the first core layer; and GP_k is the processing time for a graph process $k \in 1, \dots, N$.

Figure 6.14 shows how $T_{\text{processing}}$ values grow with increasing the data generation frequency (from 10 to 100 msg/s). Each curve is related to a different graph topology. Figure 6.15 shows how $T_{\text{processing}}$ values grow

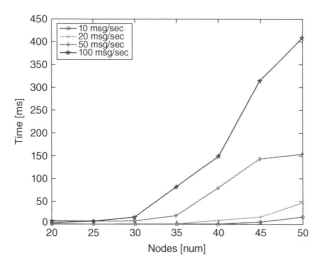

Figure 6.14 Average times related to graph framework processing block, showing per-node time, varying data generation rate, for each subset of nodes deployed into the graph topology.

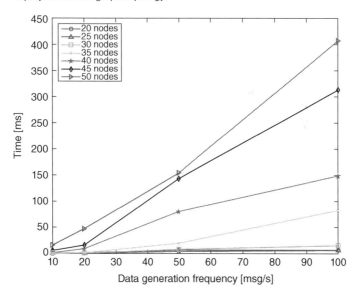

Figure 6.15 Average times related to graph framework processing block, showing per-node time, varying the subset of nodes deployed into the graph topology, for each evaluated data generation frequency.

with increasing number of nodes in the graph topology (from 20 to 50 nodes). Each curve in Figure 6.15 is related to a different value of the frequency.

6.3.5 Solutions and Security Considerations

The presented architecture is designed with reference to a specific IoT scenario with strict latency and real-time requirements, namely a smart parking scenario. There are several possible use cases and applications fitting this scenario, alerting or real-time monitoring applications.

Smart cities are encountering many difficulties in real-life deployments, even though obvious factors justify the necessity and the usefulness of making cities smarter. Vilajosana *et al.* have analyzed in detail the causes and factors that act as barriers in the process of institutionalization of smart cities, and have proposed an approach to make smart cities become a reality [208]. They advocate three different stages in order to deploy smart cities technologies and services.

- *The bootstrap phase*: This phase is dedicated to offering services and technologies that are not only of great use and genuinely improve urban living, but also offer a return on investment. The important objective of this first step is, thus, to set technological basis for the infrastructure and guarantee the system has a long life by generating cash flows for future investments.
- *The growth phase*: In this phase, the finances generated in the previous phase are used to ramp up technologies and services that require large investments; these do not necessarily produce financial gains but are only of great use for consumers.
- *The wide adoption phase*: In this third phase, collected data are made available through standardized APIs and offered by all different stakeholders to third-party developers in order to create new services. At the end of this step, the system becomes self-sustainable and might produce a new tertiary sector specifically related to services and applications generated using the underlying infrastructure.

With reference to the third phase, they propose three different business models to handle the delivery of information to third parties.

- *The app-store-like model*: developers can build their apps using a set of verified APIs after a subscription procedure that might involve a subscription fee. IoT operators can retain a small percentage of the profits of apps published in the Apple Store and/or Android marketplace.

- *The Google-Maps-like model*: the percentage fee on the sales price of an app is scaled according to the number and granularity of the queries to deployed APIs.
- *The open-data model*: this model grants access to APIs in a classical open data vision, without charging any fee to developers.

The architecture described in this paper is compatible with these steps, and, more specifically, it can adopt the Google-maps-like model, where infrastructure APIs make available different information streams with different complexity layers. The graph architecture, moreover, gives another opportunity to extend the business model, as developers can use available streams to generate a new node of the graph, and so publish a new stream for the system.

Another aspect, with a relevant impact on the business model, is security. This entails both a processing module and interaction with external entities. It is possible to adopt different policies related to authentication and/or authorization on data sources, for example based on well-known and standard solutions such as OAuth [175], preventing malicious alterations of data streams and the resulting negative consequences, which could affect both processing results and platform reliability. At a final stage, security could be applied for consumer accounting and authentication, ensuring appropriate platform access only to authenticated/authorized entities, and providing secure transactions with authorized entities via secured communications.

Security features, including authorization, authentication and confidentiality, should be integrated into the architecture, in order to make the implementation complete and usable. Details about integration of security features in the proposed big stream platform and its further impact on the system performance are not included here.

6.4 Big Stream and Security

Several methods and strategies to enable confidentiality in publish/-subscribe IoT infrastructures have been proposed. IoT systems have to avoid security threats, providing strong security foundations built on a holistic view of security for all IoT elements at all stages: from object identification to service provision; from data acquisition to stream processing. All security mechanisms must ensure resilience to attacks, data authentication, access control, and client privacy.

Collina *et al.* imagine IoT systems bridging the physical and the "virtual" worlds, using a novel broker that supports protocols such as HTTP and MQTT, adhering to the REST paradigm and allowing developers to easily and responsively expose fundamental entities as REST resources [209]. The broker does not address any security issues, the authors claiming that possible solutions could include: plain authentication; virtual private networks, access control lists, as well as OAuth, a new type of authorization used to grant third parties access to personal data (Hardt, 2012).

Lagutin *et al.* have examined the roles of different actors making up an inter-domain publish/subscribe network [210]. They consider the security requirements and minimal required trust associations between entities, introducing and analyzing an architecture that secures both data and control planes. They identify the main security goals for a publish/subscribe architecture as:

- integrity
- scalability
- availability
- prevention of unauthorized traffic.

They identify different actors and security mechanisms. The main mechanism is packet level authentication, which, combined with cryptographic signatures and data identifiers tied to secured identifiers, creates a strong binding between data and traffic, thus preventing denial of service attacks.

Wang *et al.* deal with security issues by relying on the requirements of a particular application and on an external publish/subscribe infrastructure [211]. The general security needs of the applications include confidentiality, integrity, and availability. In contrast, the security concerns of the infrastructure focus on system integrity and availability. Security issues in publish/subscribe platforms rely on authentication, information integrity, subscription integrity, service integrity, user anonymity, and information confidentiality, in addition to subscription confidentiality, publication confidentiality, and accountability.

Raiciu and Rosenblum have presented a study of confidentiality in content-based publish/subscribe (CBPS) systems [212], defined as an interaction model storing the interests of subscribers in a content-based infrastructure, to guide routing of notifications to subjects of interest. In agreement with the Wang *et al.* approach [211], confidentiality aspects are decoupled into two facets, namely notification and subscription, suggesting that a confidential CBPS (C-CBPS) must satisfy correctness and notification, whereas a subscription CBPS must satisfy unforgeability and security, and match isolation. A high-level approach to obtain C-CBPS by relying on notifications using simple blocks that may be controlled and checked more easily than if they were completely encrypted, is proposed.

Fremantle *et al.* have analyzed the use of federated identity and access management in the IoT [213]. They follow a consumer-oriented approach in which consumers own data collected by their devices, and have control over the entities that access these data. Traditional security models, based on the concept of roles in a hierarchical structure, are not applicable for IoT scenarios (because of the billions of devices involved and thus the impossibility of adopting a centralized model of authentication, and the necessity to support mechanisms for delegation of authority). The authors therefore proposed OAuth2 [175] as a possible solution that could achieve access management for IoT devices that support the MQTT protocol. The overall system consists of:

- a MQTT broker
- an authorization server supporting OAuth2
- a web authorization tool
- a device.

Bacon *et al.* tackle the problem of application security in the cloud, aiming at incorporating end-to-end security so that cloud providers can not only isolate their clients from each other, but can also isolate the data generated by multiple users who access a particular service provided by the cloud [214]. They propose an approach called "application-level virtualization," which consists of:

- removing from applications all details regarding security and flow control;
- placing the security management logic in the cloud;
- allowing providers to permit only those interactions that the clients specify,

6.4.1 Graph-based Cloud System Security

Addressing the security problem in a graph-based cloud system requires a broad approach, owing to different needs of the components involved. In Figure 6.16, the main building blocks listed above are shown. The main components, and their respective security mechanisms, are indicated.

The enhanced graph architecture provides security by means of two modules:

- *The outdoor front-end security module* (OFS) carries out security operations that can be applied a priori, before receiving data from a generic external source, as well as before a final consumer can start interacting with the graph-based cloud platform.
- *The in-graph security module* (IGS) adopts security filters that can be applied inside the heart of the graph-based cloud platform, so that processing nodes are able to control access to the streams generated by internal computational modules.

The OFS module is crucial for infrastructure safety: its role includes monitoring access to the platform and authorizing information flows coming from, or directed to, external entities. On one side, OFS must verify and authorize only desired external input data sources, allowing them to publish raw streams in the IoT system. On the other hand, OFS is required to secure outgoing streams, generated by different layers of the graph platform itself, authorizing external consumers to use, them.

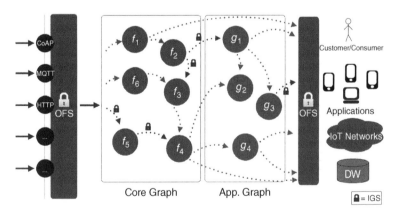

Figure 6.16 Main building blocks of the proposed listener-based graph architecture. Nodes in the graph are listeners, edges between nodes represent dynamic flows followed by information streams. Edges can be "open" or "secured".

Consider, as example, the case of company C, which owns a set of particular sensors, and wishes to become an IoT stream source for the graph-based cloud platform. C wants:

- to sell sensed data only to a specific subset of customers, in order to protect its commercial interests
- to make a profit from these sales.

Therefore, the OFS module is strictly related to sensors and devices at the input side, and to customers' smart objects at the output stage, so that it becomes protocol-dependent and can be adapted to the specific technologies supported by the target devices.

The IGS module is not related to the OFS module, as it acts exclusively at the heart of the IoT graph architecture, coordinating and managing inner inter-node interactions. The IGS module must be implemented inside single processing nodes, enabling them to define a set of rules that describe what entities may become listeners of a generated stream.

Referring to the graph architecture, shown in Figure 6.16, edges in the graph can be classified as follows:

- "Open" edges are data streams generated by core or application nodes in the graph platform, which can be forwarded to all interested listeners without need for isolation or access control.
- "Secured" edges are data streams that should comply with specified rules or restrictions that determined the possible consumers of the generated data.

As an example, consider again company C, which provides its sensors as data sources and has notified the architecture that the streams produced by its sensors should be secured.

The integration of security modules in the IoT architecture entails modifications in the structure and the modules of the (unsecured) architecture previously described.

In the rest of this section, an analysis of each module in the graph architecture is presented, in order to explain how security mechanisms can be embedded and managed. In particular, we introduce the OFS module, which supports the acquisition and normalization modules on authorization of external entities. In addition, an enhanced version of the application register is described in order to underline the management of secure interactions with processing nodes. Finally, an overview inside the graph nodes, analyzing how security is applied in the processing stages, is presented.

6.4.2 Normalization after a Secure Stream Acquisition with OFS Module

The acquisition and normalization modules, shown in Figure 6.17, represent the entry point for external sources (e.g., smart objects deployed in different IoT networks) to the proposed architecture.

The purpose of the acquisition block is to receive incoming raw data from heterogeneous sources, making them available to all subsequent functional blocks. Since raw data are generally application- and subject-dependent, the normalization block has to "normalize" incoming data, generating a common representation, suitable for further processing. This might involve suppression of unnecessary data, data enrichment, and format translation.

After the normalization process, data are sent to the first core layer. Within each graph layer, streams are routed by dedicated components, called brokers. These are layer-specific and, in the current implementation of the architecture, are RabbitMQ exchanges. As stated before, smart objects can communicate using different protocols. For this reason, the acquisition block has to include a set of connectors, one for each supported protocol, in order to properly handle each protocol-specific incoming data stream.

As shown in Figure 6.17, these modules must cooperate with the OFS module, which has to be activated before an external source is able to operate with the graph platform. At the acquisition stage, in order

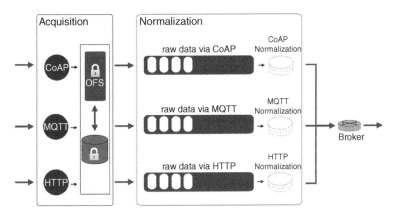

Figure 6.17 The OFS module manages security in the acquisition and normalization blocks, interacting with an authentication storage entity containing data source identities.

for the proposed IoT platform to support both "open" and "secured" communications, protocol-specific communication security mechanisms have to be implemented at the relevant layers: at the network layer through IPSec, at the transport layer through TLS/DTLS, at the application layer through S/MIME or OAuth.

As stated before, the current implementation supports different application protocols at the acquisition stage, namely MQTT, HTTP, and CoAP. In order to secure all communications with these protocols, we need to introduce different protocol-specific policies. Fremantle *et al.* proposed an OAuth-based secure version of the MQTT protocol [213], showing that MQTT also complies with an n-legged OAuth protocol.

The proposed IoT platform provides a good way to authenticate external data providers, adopting open-source and well-understood solutions. The OFS module can be secured by OAuth, used in ways dependent on the specific communication protocols supported by the heterogeneous IoT smart objects.

A suitable solution to provide authorization in IoT scenarios is IoT-OAS, as presented in Section 5.3.2, which represents an authorization framework to secure HTTP/CoAP services. The IoT-OAS approach invokes an external OAuth-based authorization service (OAS). This approach is meant to be flexible, highly configurable, and easy to integrate with existing services, guaranteeing:

- lower processing loads than solutions with access control implemented in the smart object;
- fine-grained (remote) customization of access policies;
- scalability, without the need to operate directly on the device.

Returning to the previous example, company C, to became a secured IoT data source, selects one of the supported protocols (HTTP, CoAP or MQTT) to send raw data streams in the secured version.

6.4.3 Enhancing the Application Register with the IGS Module

One of the main motivations for securing a system internally is the need to secure some of its operations, as well as to isolate some processing steps of the entire stream management. The security features should be coordinated by the application register module, which maintains and manages interactions between inner graph nodes of the IoT

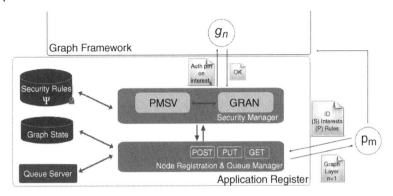

Figure 6.18 The application register module structure with security elements. PMSV and GRAN modules interact with a storage entity to manage authorization in the graph.

platform using different communication protocols, as requested by the architecture itself.

In order to accomplish the operational functionalities listed previously, the application register module has two main components, as shown in Figure 6.18.

- *The graph state database* is responsible to maintain all information about the current graph status. Since this component is not critical from a performance viewpoint, it has been implemented through a simple relational SQL database.
- *The node registration and queue manager* (NRQM) *module* is responsible for managing communications with existing graph nodes, as well as with external entities that ask to join the big stream architecture.

To add security features to these modules, the application register defines entities and modules specifically related to security management and coordination. As shown in Figure 6.18, the application register comprises the following additional modules:

- The *policy manager and storage validator* (PMSV) is responsible for managing and verifying authorization rules, interacting with the persistent storage element, which maintains authorization policies.
- The *graph rule authorization notifier* (GRAN) interacts with publisher graph nodes and verifies if listener nodes are authorized to receive streams associated with specific topics.

- A *persistent storage entity* (e.g., a non-relational database) maintains authorization rules specified by publisher graph nodes.

If a processing node p_m asks to register with the IoT platform, requiring it to join the graph, after authentication (e.g., using a username/password pair, cryptographic certificates, ACLs, OAuth) there are two cases that require the use of security mechanisms and involve the defined modules:

- a registration request coming from a node that is willing to become a publisher node for a secured stream (e.g, an application node created by developers of company C);
- a registration request sent by a node, asking to be attached as a listener for some streams.

In the first case, when an external process p_m requests to register to the graph architecture, in order to secure one or more of its own streams, it updates the PMSV module. After indicating its published topics, it specifies some policies and rules, to be stored, together with the assigned operative graph layer, in the persistent security storage by PMSV itself. These rules will be checked in case of future subscription requests for the node.

In the second case, an external process p_m, upon issuing a request to attach to the graph platform and to become a node, provides information related to its identity and also specifies the interests to which it wants to subscribe. The application register, having identified the graph layer into which the new node could be placed, takes charge of these interest specifications, passing then to the PMSV, which acts as follows.

For each provided *interest$_k$*, the PMSV module interacts with the persistent storage entity, performing a lookup for a match between the interest and stored publishing policies, and refining this lookup with layer matching:

$$match = \{layer = x \quad OR \quad layer = (x+1)\} \quad AND$$
$$\{interest_k \in \Psi\} \tag{6.3}$$

where x stands for the identified listener graph layer; indicates each single specified interest, extracted from the attaching request; represents the persistent storage element; and contains a list of publisher nodes that have to authorize subscriptions.

If *match* contains some positive results (e.g., g_m node – a node in the graph), these are forwarded to the GRAN module, which interacts with discovered publisher nodes, sending them the identity of the requesting listener node and asking them to allow or deny subscription to the requested topics. This response is sent back to the GRAN module, which analyzes it and, in compliance with the application register, authorizes or rejects the listener graph node subscription.

In order to better explain the behavior of the application register module in relation to the join operation of an external entity that asks to become a graph listener, in Listing 6.3 we detail the interactions involved through a pseudocode representation.

Listing 6.3 Pseudocode representation of the operations of the application register when an external process asks to become a graph listener in the big stream architecture.

```
NodeIdentityCertificate = {
  NodeInfo = {....};
  INTERESTS = {interest_1, interest_2, ..., interest_N};
}
MAIN() {
 Pm = receive_join_request(NodeIdentityCertificate);
 authenticated = authenticate_node(Pm);
 if (authenticated is TRUE) {
    x = identify_graph_layer_for_node(Pm);
    foreach (interest interest_k in INTERESTS) {
      Match = send_request_to_PMSV(layer = (x OR x+1),
    interest = interest_k);
      if (Match is EMPTY) {
        REPLY_DENY(topic = interest_k,
          destination_node = Pm);
      }
      else {
        foreach (GraphNode Gm in Match) {
          allow = request_grant_to_owner_via_GRAN(topic =
    interest_k, owner = Gm, listener = Pm);
          if (allow is FALSE) {
            REPLY_DENY(topic = interest_k,
              destination_node = Pm);
          }
          else {
            REPLY_SUCCESS (topic = interest_k,
              destination_node = Pm);
          }
        }
      }
    }
  }
}
```

```
REPLY_DENY(topic, destination_node) {
  send_DENY_response_to_destination(required topic = topic);
}
REPLY_SUCCESS(topic, destination node) {
  send_SUCCESS_response_to_destination(required topic =
    topic, security_parameters = (....));
}
```

The processing nodes in the graph architecture can be both listeners and publishers at the same time, so that the previously detailed mechanisms can be applied together, without any constraint on the execution order. The flows shown in Figure 6.18 represent the interactions in this mixed case. The rule on node authority, restricted to the same layer and to the next one, decreases lookup times in rule matching execution.

Moreover, external smart object producers could also request a totally "secured" path, from source to final consumer. These constraints have a higher priority than policies defined by publisher graph nodes, being specified by the stream generators. In this way, these external priority rules are stored in the persistent storage elements as well, and when a new graph node registers to the proposed IoT platform, its graph-related policies are left out, and it is forced to comply with the external rules.

6.4.4 Securing Streams inside Graph Nodes

The graph framework comprises several processing entities, which perform various computations on incoming data, and each representing a single node in the graph topology. The connection of multiple listeners across all processing units defines the routing path of data streams, from producers to consumers. All nodes in the graph can be input listeners for incoming data and output producers for successor graph nodes.

Since each graph node is the owner of the stream generated by its processing activity, it is reasonable to assume that it could decide to maintain its generated stream "open" and accessible to all its interested listeners, or to applying security policies, isolating its own information and defining a set of rules that restrict the number of authorized listener nodes. In this latter case, a "secure" stream is created and encrypted using algorithms selected by the owner. Each listener is thus required to decrypt incoming data before performing any processing. These encryption/decryption operations can be

Graph Framework

Broker

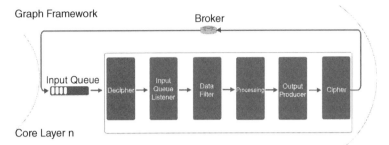

Core Layer n

Application Layer m

Figure 6.19 Detail of the structure of a single graph node: the decryption module is activated when the node is a listener of a secured stream, while the encryption module is activated if the node generates a secured stream.

avoided if listeners adopt homomorphic encryption [], allowing them to carry out computations on ciphertext, instead of on plaintext. This generates an encrypted result that matches one performed on the plaintext, but without exposing the data in each of different steps chained together in the workflow. Homomorphic encryption allows computation to be executed in the encrypted domain, thus providing end-to-end security, and avoiding the need for hop-by-hop encryption/decryption.

In Figure 6.19, the modules inside a graph node are shown: the broker of the core layer forwards streams to interested graph nodes, placing these data in the input queues of each. The output stream generated by the processing of these nodes, will be sent to the same broker in the core layer, which is linked to the broker of the next graph layer, and which "spreads" generated streams to all interested nodes. Some of these modules will be activated only in specific situations. In particular, the node illustrated acts as a listener of a "secured" data stream, so it has to decrypt an incoming message, activating the decryption module. It also acts as a producer of a "secure" stream and so it has to encrypt its processed streams with the encryption module before forwarding it, thus hiding the stream from unauthorized listener graph nodes. This is the case in the earlier example in which a graph node that is already a listener of the secured stream owned by company C, wants to secure the stream generated by its processing.

It is important to point out that each graph node controls its generated flow, with visibility of only one step. This means that a listener of a "secured" flow can publish an "open" stream and vice versa, thus producing "hybrid" path combinations. These, in the stream flow from

Table 6.1 Comparison between graph framework actors and OAuth roles.

Graph framework actor	OAuth role
Publisher graph node, owner of the outgoing data stream	Resource owner
Listener graph node, willing to subscribe to interested topics	Consumer
Infrastructure routing element (broker in a pub/sub paradigm)	Provider

IoT source to final consumer, produce a combination of "secured" and "open" steps.

Referring to the example in which company C generates a "secured" stream of data coming from its sensors, an IoT developer might decide to create a new graph node listening to both the secured stream of company C and the stream of another company, D. In the processing unit of the new graph node, the developer can aggregate and transform the input streams, generating new and different output streams, which can be published in "open" mode, since the developer is the owner of this new produced stream. Depending on the inner organization of the IoT architecture, there could be a parallel between actors enrolled in the graph framework and OAuth roles, as illustrated in Table 6.1.

More precisely, OAuth roles could be detailed as follows.

- *Resource owner*: the entity that owns the required resource and has to authorize an application to access it, according to authorization granted (e.g., read/write permission).
- *Consumer*:the entity that wants to access and use the required resource, operating in compliance with granted policies related to this resource.
- *Provider*: the entity that hosts the protected resource and verifies the identity of the consumer that issues an access request to this resource.

As previously stated, each graph node can apply cryptography to its streams, using encryption and decryption modules. The security mechanisms leave a few degrees of freedom to developers, who can adopt their own solutions (e.g., using OAuth tokens) to secure streams, or rely on standard secure protocols, thus adopting well-known and verified solutions. An overall view of the envisioned IoT architecture is shown in Figure 6.20, showing all component modules and their interactions.

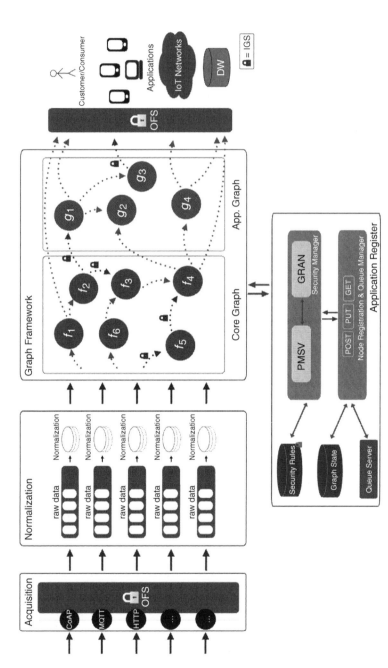

Figure 6.20 Complete IoT cloud architecture, including proposed security modules and showing different interactions, from incoming stage to final consumer notification.

6.4.5   Evaluation of a Secure Big Stream Architecture

The evaluation involves a Java-based data generator, which:

- simulates events arrivals from IoT sensors networks;
- randomly selects an available protocol (HTTP, CoAP, MQTT);
- periodically sends streams to the corresponding interface in the acquisition module.

Once received, data are forwarded to the dedicated normalization module, which enriches them with parking membership information from an external SQL database. The module structures the stream in a JSON schema compatible with the architecture. Once it has completed its processing, it sends the structured data to the graph framework, which forwards the stream following paths based on routing keys, until the final external listener is reached. The graph considered in our experimental set-up comprises eight core layers and seven application layers, within which different graph topologies (from 20 to 50 nodes) are built and evaluated.

The proposed architecture was evaluated by varying the incoming data stream generation rate between 10 msg/s and 100 msg/s. The first evaluation, which represents a benchmark for our performance analysis, uses the platform without security mechanisms. Then, security mechanisms were introduced into the graph framework module, in order to assess the impact of a security stage on the overall architecture.

The first performance evaluation was conducted by measuring the delay between the time when normalized data were injected into the graph framework and the time instant at which the message reached the end of its routing, becoming available for external consumers/customers. In order to consider only the effective overhead introduced by the architecture, and without taking into account implementation-specific contributions, performance results were obtained by subtracting the processing time of all core and application nodes. Finally, these times were normalized over the number of computational nodes, in order to obtain the per-node overhead introduced by the architecture, in a way that is independent of the implemented routing and topology configuration.

The second performance evaluation adopted the same structure as the unsecured implementation, but introduced security mechanisms inside graph nodes, through the adoption of symmetric encryption to encrypt/decrypt operations, as described in Section 6.4.4. In order to guarantee a tradeoff between security-level and reliability, we chose the Advanced Encryption Standard (AES), which is a symmetric cryptosystem, in its 256-bit key version [117]. AES is a block cipher based on a substitution and permutation (SP) combination, working on 128-bit blocks. The chosen key size determines the number of repetitions of transformation rounds that convert the input, applying several processing stages and depending on the encryption key. The strength of AES256 derives from its key space – a possible 256-bit key – which affects the time needed to made a successful brute-force attack on a system in which it is implemented.

In the second evaluation, we also implemented a new version of the processing core and application nodes, applying security at both input and output stages of the single graph node, with the following behavior:

- If the processing node has received an AES 256-bit decryption key from the application tegister, it then uses this key to decrypt incoming messages, returning plaintext useful for processing operations.
- If an AES 256-bit encryption key was provided by the application register to a graph node, it then encrypts the processed stream using this key, before forwarding the stream to the relevant exchange.

This security model is also applicable also to the example in which company C would like to secure its paths into the graph framework. In the second evaluation, encryption and decryption keys were provided to all graph nodes, in order to secure all the intermediate steps followed by streams owned by the company.

Different topologies were obtained by varying the subset of nodes deployed, from 20 to 50, and the data generation rate, from 10 msg/s to 100 msg/s. The results are shown in Figure 6.21.

The stream delay can be given using the following expression:

$$T_{\text{processing}_{\text{freq}}} = \frac{T_{\text{out}} - T_{\text{in}} - \sum_{k=1}^{N} GP_k}{N} \tag{6.4}$$

where T_{out} is the instant at which parking data reach the last application processing node; T_{in} indicates the instant at which normalized data comes to the first core layer; and GP_k is the processing time of a graph node $k \in \{1, \dots, N\}$.

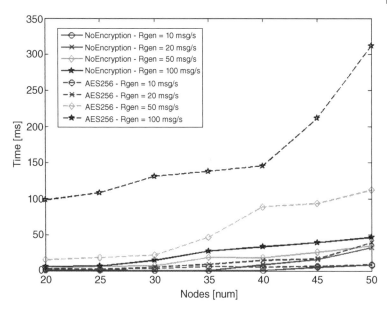

Figure 6.21 Average stream delay (ms) related to graph framework processing block, showing per-node time, in the case of unsecured communication as well as the case of adoption of symmetric encryption.

In order to investigate the benefits and drawbacks of security solutions other than symmetric encryption, we implemented an asymmetric-cryptography version of the graph processing nodes, adopting RSA [132] with a 512-bit key. This is a private-public key cryptosystem. In a third evaluation scenario, the symmetric cryptosystem was replaced with private/public RSA certificates provided to the graph nodes by the application register module.

The results, in terms of stream delay, are shown in Table 6.2. They highlight that an asymmetric cryptosystem is a bad choice for graph inter-node security. Asymmetric solutions might be adopted outside of the graph nodes, when an external node is willing to become an operating entity of the graph framework, challenging an authentication transaction with its signed certificate, which allows for verification of its identity by the application register. Therefore, in the joining phase, asymmetric solutions could be used if time is not the main constraint.

Table 6.2 Average stream delay related to the adoption of asymmetric encryption solution (RSA) into the graph framework processing block.

Number of nodes	Stream delay (ms)	
Message rate (msg/s)	50	100
20	128	10890
25	156	12962
30	2783	13841
35	10104	14048
40	11283	14515

In order to better highlight this final analysis of the evaluation results, in Figure 6.22 a logarithmic-scaled version of the results is shown, evaluating the logarithm of the stream delay as a function of the number of nodes in the graph for:

- no encryption,
- symmetric encryption (AES256)
- asymmetric encryption (RSA512).

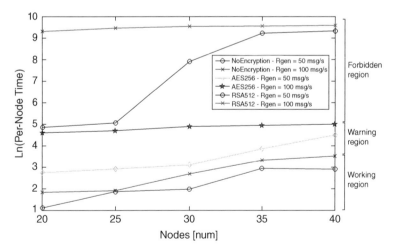

Figure 6.22 Logarithmic representation of the stream delay as a function of the number of nodes of the graph, evaluated both with and without security mechanisms.

Two values of data generation rate were used: 50 and 100 msg/sec. The results suggest that there are three main performance regions in which the proposed big stream platform could be considered:

- the working region, approximately around the "NoEncryption" curves, for which the system has the benchmark results, and where processing does not introduce heavy delays;
- the "Warning" region, around the AES256 curves, in which delays introduced by security steps degrade performance a little, while maintaining good quality of service (QoS);
- the "Forbidden" region, around the RSA512 curves, in which the system incurs major delays, which may cause crashes and dropping of service, invalidating QoS and any service level agreements (SLAs) signed with data stream producers and consumers.

6.5 Fog Computing and the IoT

The role of cloud computing in the IoT is gaining more and more attention. Most research has so far been focused on smart objects, and in particular on the definition of efficient, lower-power, IP-based communication protocols and mechanisms. Many of these areas have now been significantly addressed. IoT solutions have been deployed and brought to the market in several application scenarios, from home automation to smart cities. Most of these fragmented and vertical solutions rely on the cloud to provide centralized access to services exploiting data that are sent uplink from deployed sensors to cloud storage. Typically, mobile apps "consuming" such services are made available to end-users. However, this approach, which is useful for the IoT, does not fully exploit the potential of the cloud.

Fog computing, also referred to as "Edge Computing", is a novel paradigm that was introduced to meet requirements such as mobility support, low latency, and location awareness [205]. Fog computing aims to moving some cloud-based computation and storage to the edge of the network. The fog is a cloud close to the ground and, as such, provides end users with computing functionality that is closer to them, thus improving performance. It brings low-latency to consumers and enables development of new applications that take into account location-related information.

The characteristic features of fog computing are the following:

- wide geographical distribution, in contrast to the centralization envisioned with the cloud;
- subscriber model used by the players in the fog;
- support for mobility.

Fog computing brings a new approach to Internet access networks by making computation, storage, and networking resources available at the edge of access networks. This improves performance, minimizing latency and maximizing availability, since resources are accessible even if Internet access is not available [215].

Fog-based solutions aim at introducing an intermediate architectural layer in which resources and applications are made available in the proximity of end devices, thus avoiding the need for continuous access to the cloud. While cloud-only architectures can provide a solution to scalability and flexibility issues by distributing resources among multiple servers, this approach has some weaknesses, such as:

- latency
- availability/dependence on Internet connectivity for operations
- lack of flexible networking
- quality of service/experience
- security and privacy.

Due to its benefits over cloud-based architectures, especially if time is a critical issue or Internet connectivity is poor or absent, fog computing is expected to play a key role in the deployment of IoT applications. The fog is not intended to replace the cloud, but rather to complement it, in order to provide location-aware and real-time services, thus enabling new applications that could have not been deployed otherwise.

Fog-based access networks are based on the presence of highly specialized nodes, called fog nodes, which are able to run distributed applications at the edge of the network. In particular, the deployment of computing resources on Internet access networks allows for dynamic activation of Virtual Machines (VMs) on fog nodes. For this reason, the cloning and synchronization techniques of VMs at the core of this work fit perfectly into Fog-based infrastructures, as will be discussed in more detail in the following sections.

The proposed architecture can protect local resources by providing remote access to their replicas in a transparent way. Local resources are kept synchronized by multiple clones of the same machine, thus achieving a high level of reliability and load balancing. Smart management of the activation/deactivation of replicas and the choice of the most appropriate fog node to run the clone allows for optimization of the usage of CPU and memory on the infrastructure, according to the specific real-time resource requirements of the applications involved.

A lightweight alternative to VMs are containers, which provide a more flexible environment for "disposable applications" like the IoT Hub. Container platforms like Docker [216] are gaining increasing attention for fog computing applications. Moving from a centralized to a decentralized paradigm enables processing to be offloaded to the edge, reducing application response times and improving overall user experience. This process will play a fundamental role in IoT.

Several authors have described how a container-based architecture could be efficiently used for dynamic networking applications [217, 218]. Ramalho and Neto have compared existing lightweight and hypervisor-based approaches for edge computing and presented an efficient networking approach [219]. Morabito has presented a novel approach to the application of a lightweight virtualization technology (such as Docker) to constrained devices with a negligible overhead [220].

In this work a containerized version of the IoT Hub, based on Docker, will be considered.

6.6 The Role of the IoT Hub

The billions of IoT sensing/actuating devices in the IoT will generate an unprecedented amount of data, which will need to be managed properly in order to provide highly available and robust services. Although an IP-based IoT would make it possible to address smart objects directly, the following potential drawbacks exist:

- In order to extend their battery lifetimes, smart objects may be duty-cycled and thus not always accessible.
- Smart objects might be unable to handle a large number of concurrent requests, thus leading to service disruption and becoming possible targets for denial-of-service attacks.
- In some circumstances (for instance, to minimize memory footprint when the available in-device memory is critically low), smart objects may play the role of clients rather than servers.

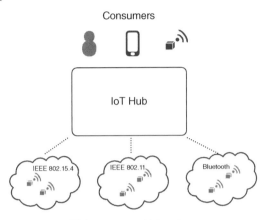

Consumers

IoT Hub

IEEE 802.15.4   IEEE 802.11   Bluetooth

Heterogeneous Networks

Figure 6.23 The IoT nub can manage multiple networks of heterogeneous smart objects and enables access to resources by external consumers that should not be aware of the ow-level details of communications.

In all the above cases, the presence of an intermediate network element operating at the application layer – typically the border router – is desirable. Such a node could integrate several functionalities that would reduce the processing load on smart objects and help overcome some of the problems discussed above (e.g., caching). This resourceful node, called the IoT Hub is shown in Figure 6.23. It helps move some of the processing load towards the edge of the network, following the fog computing paradigm [205]. Even though this approach will surely bring several benefits, the role of the IoT Hub then becomes central in the IoT architecture foreseen, since it will be responsible for processing all requests.

As billions of smart objects are expected to be deployed, efficient data processing has led to reliance on the cloud. The expression "Cloud of Things" is sometimes used to refer to the interaction between IoT and the cloud [221]. Aazam *et al.* have set out an architecture for integrating the cloud and the IoT is proposed, based on a network element called a smart gateway [222]. This is intended to act as intermediary between heterogeneous networks and the cloud. The role of the smart gateway is similar to that of the IoT Hub, in terms of supporting several heterogeneous networks. However, the role of the cloud is mainly envisioned as being for data aggregation and storage, and usable by end-users to access data. In this approach, data are sent uplink, making it impossible to directly address and act on response from smart objects, as is supposed to happen in the IoT.

At the opposite extreme, in the current paper we envision that the cloud, by hosting replicas of the IoT Hub, is used to give direct and efficient access to resources, while providing desirable features such as seamless access by external clients, security, and high availability.

6.6.1 Virtualization and Replication

In order to increase the robustness of an hub-oriented IoT architecture, we propose a cloud-supported replication mechanism for IoT Hubs in order to efficiently manage CoAP resources in a scalable and secure way. The proposed mechanism uses cloud platforms to clone and virtualize IoT Hubs. The replicas are full copies of the IoT Hubs and so implement all their functionalities. Accessing the replicas is therefore like asking a delegate for the same information. This brings several benefits:

- A unique cloud-based interface for accessing resources is exposed.
- The actual implementation details of the IoT Hub are hidden from communications, thus protecting the IoT Hub and the smart objects behind.
- The cloud platform may introduce balancing policies in order to scale up or down the number of replicas according to current needs and the number of incoming/estimated requests and resources to be managed.

This solution enables remote access to resources in networks in a fully transparent and standardized way. In order to achieve our goals, a new application layer for IoT networks is designed and developed. In particular, we focus on the design of an architecture that allows access by external clients, by virtualizing the functionalities of an IoT network. The details of the IoT network, such as its location or its actual implementation, are kept hidden from users and external clients seeking to access resources. In other words, resource access by remote clients will be mediated by the cloud platform, which provides a standard and secure front end.

6.6.1.1 The IoT Hub

The IoT Hub does not have the same strict requirements on energy consumption and processing capabilities as other smart objects and it is thus useful to provide relevant features to a constrained network. The IoT Hub is placed at the edge of the constrained network and plays

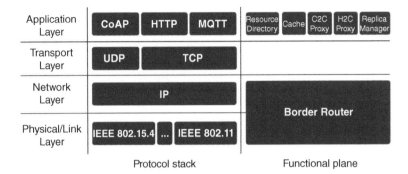

Figure 6.24 Protocol stack and functional modules implemented by an IoT Hub.

a fundamental role by implementing the functions – summarized in the functional plane and mapped in the protocol stack – shown in Figure 6.24.

- *LoWPAN border router*: at the network layer, the IoT Hub is the gateway between one or more constrained networks (e.g., IEEE 802.15.4) to which it belongs (via its radio interfaces).
- *CoAP/CoAP (C2C) proxy*: at the application layer, this provides proxying capabilities for CoAP requests coming from external clients that should reach internal constrained nodes.
- *HTTP/CoAP (H2C) proxy*: at the application layer, this provides cross-proxying (protocol translation) between HTTP and CoAP in order to let external HTTP clients access CoAP resources hosted by smart objects.
- *Resource directory*: the IoT Hub maintains a list of all CoAP resources available in the constrained network. These resources may have been gathered through several mechanisms, such as those described in Chapter 4.
- *Cache*: in order to avoid unnecessary load on smart objects and to minimize latency, a cache is kept with a representation of the most recently accessed resources.
- *Replica manager*: this is a software module responsible for the coordination of and synchronization between the IoT Hub and its replicas.

Due to the constrained nature of smart objects, they cannot typically implement strong security mechanisms and access policies, for instance those related to authorization, in which the IoT Hub may act as a filter for incoming requests in order to limit access to specific resources. Because of all the functionalities outlined above, the IoT Hub may suffer from the following critical issues, which may undermine the lifecycle of a constrained IoT network:

- The IoT Hub is a bottleneck of the architecture, since all traffic must pass through it even if it is not a communication endpoint.
- Failure of the IoT Hub would make the resources hosted by smart objects temporarily or permanently unavailable.

It is necessary to relieve the IoT Hub from some of this load in order to guarantee that resources can be accessed with high availability in a secure and seamless way.

In this section, we propose an approach that relies on on the cloud to provide virtual replicas of the IoT Hub. Replicas of the IoT Hub are fully functional clones, which may be used by any external client to access resources in the same way as they would do with the actual IoT Hub. Interacting with resources through the replicas gives an access point that is different from the real (physical) IoT Hub, thus decoupling the constrained network management function and granting access to external clients.

Replicas of the IoT Hub are synchronized through a dedicated MQTT-based protocol, which is used to transfer copies of the resources from the IoT Hub to the replicas in a pub/sub model. Replicas can be instantiated on the fly, according to particular needs, thus also acting as load balancers (according to policies which may depend on the number of connected clients and smart objects) and as recovery facilities in the event of temporary failure of the real IoT Hub.

6.6.1.2 Operational Scenarios

Resource access through the proposed cloud-based platform can occur according to the following three operational models implemented by a smart object:

- CoAP server
- observable CoAP server
- CoAP client.

Polling Resources

If the smart object is a CoAP server, it can receive requests to access its hosted resources. In this case, as shown in Figure 6.25, the message

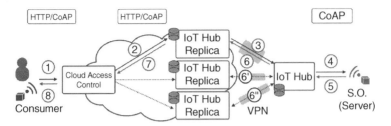

Figure 6.25 Message flow for the polling scenario: a HTTP/CoAP client requests a resource to the cloud platform, which internally selects a suitable IoT Hub replica and forwards the request. The request reaches the smart object only if neither the replica nor the IoT Hub have stored a fresh cached representation of the resource. Other replicas (which are not in the path of the request) are kept in sync using the synchronization protocol.

flow is as follows: (1) the external client sends a HTTP or CoAP request to the cloud platform front-end, which (2) forwards the request to one selected replica of the IoT Hub. If the replica has a "fresh" matching cached resource, it can return this immediately; otherwise (3) the replica forwards the request to the actual IoT Hub (which is securely connected through a VPN tunnel). If the IoT Hub has a fresh matching cached resource, it can return it immediately; otherwise (4) it acts as a reverse proxy and forwards a (and, if needed, translated) CoAP request to the CoAP server, which (5) returns the resource. At this point, the returned resource is cached by the IoT Hub, and (6) returned to the replica, which, in turn, (7–8) sends it back to the client. The resource cached at the IoT Hub is then (6' and 6'') synchronized with its replicas, in order to speed up and efficiently manage subsequent requests targeting the same resource.

Observing Resources

The Observe option [9] is a CoAP option that allows resource observing. According to this specification, a CoAP client can send a single request, including an Observe option with a value of 0 (register), in order to register its interest in receiving updates related to the targeted resource. The CoAP server sends a response each time the resource value is updated. In this case, multiple responses are sent after a single request. The Observe option allows a notification-based communication model to be implemented, thus reducing network traffic. The observing scenario is shown in Figure 6.26. If the targeted smart object is a CoAP server which implements the Observe option, it can receive

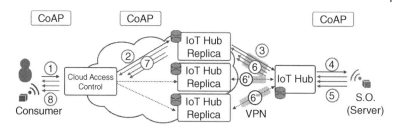

Figure 6.26 Message flow for the observing scenario: a CoAP client requests a resource to the cloud platform using the CoAP Observe option. The cloud platform internally selects a suitable IoT Hub replica and forwards the request. The observe request is then forwarded to the IoT Hub and to the smart object thus creating an "observe chain". Resource updates are then sent from the smart object back to the IoT Hub, then to the replica, and finally to the CoAP client. Other replicas (which are not in the path of the request) are kept in sync using the synchronization protocol.

requests to access its hosted resources, which may be either observable or not. In the latter case, requests are handled as in the polling case. In the former case, instead, the message flow resembles that of the polling scenario, but:

- the IoT Hub observes the resource
- the external client observes the cached resource on the replica.

When the resource is updated:

- the smart object will send a notification to the IoT Hub;
- the IoT Hub will synchronize the resource with its replica;
- the replica will send a notification to the external observing client.

Pushing Resources

Sometimes, the memory constraints of smart objects make it unfeasible to let them act as CoAP servers. In this case, as shown in Figure 6.27, the smart object acts as a CoAP client and sends CoAP requests (POST and PUT) to the IoT Hub, which plays the role of an origin server, maintaining resources on behalf of the clients. If the smart object is a CoAP client, according to the semantics of CoAP methods, the following message flow takes place: (1) the smart object sends CoAP POST requests to the IoT Hub in order to create resources and CoAP PUT requests in order to change their value. When the IoT Hub receives POST and PUT requests, after handling these requests as necessary, it stores them and (2, 2' and 2") synchronizes them

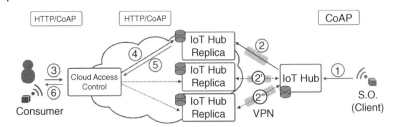

Figure 6.27 Message flow for the pushing scenario: a smart object acting as a CoAP client posts and updates resources on the IoT Hub, which acts as origin server. All replicas are kept in sync using the synchronization protocol. External client can request resources, which will served by a replica.

on the replica. When (3) an external client sends an HTTP or a CoAP request to the cloud platform front end, (4) the latter forwards the request to one selected replica of the IoT Hub. Since the replica is synchronized with the IoT Hub, (5–6) the replica can respond immediately with the stored representation of the resource.

6.6.1.3 Synchronization Protocol

The synchronization protocol used in the architecture implements a pub/sub communication model. Communication follows a one-to-many pattern from the IoT Hub to all of its replicas. All messages are sent by the IoT Hub to an MQTT message broker (hosted on the cloud platform – see Figure 6.28a). Specific MQTT topics can be used to selectively target one, many, or all replicas in order to implement unicast, multicast, or broadcast communications respectively. The MQTT broker is managed by the cloud platform, using a VPN connection, for security and addressing reasons, as shown in Figure 6.28a. The IoT Hub and its replicas are all identified by a system-wide identifier, assigned by the designed cloud platform.

Each IoT Hub includes a replica manager (RM), which is a dedicated software module responsible for the synchronization of the IoT Hub and its replicas. The RM comprises the following items, as shown in Figure 6.28b:

- a *replica registry* (RR), which contains the list of the identifiers of all the replicas of the IoT Hub;
- an *MQTT subscriber*, which registers to the broker to receive messages related to two topics (which may coincide in the case of the actual IoT Hub):

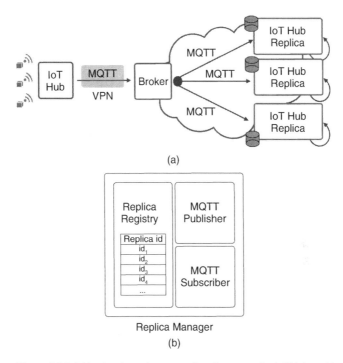

(a)

(b)

Figure 6.28 (a) broker-based message flow between the IoT Hub and its replicas. (b) internal structure of the replica registry module of the IoT Hub.

- – its own identifier (id_i)
- – the identifier of the actual IoT Hub (id_{hub});
- • an *MQTT publisher*, which publishes messages to the broker, using the method $pub(t, m)$, where t is the topic and m is the message to be published.

The IoT Hub is in charge of maintaining full synchronization of its resources with its replicas, in order to ensure that all requests are served in the same way, regardless of the specific replica that was targeted by the client. Synchronization comes into play every time a resource on the IoT Hub changes. This can be caused by different events. At startup, a replica of the IoT Hub needs to synchronize with the actual IoT Hub. The procedure is shown in Figure 6.31. The RM of the replica publishes its id_i to the topic id_{hub}, in order to notify of its creation ($pub(id_{hub}, id_i)$). At this point, the IoT Hub updates its RR by adding id_i and then starts publishing to the broker all the resources (R)

using the topic id_i ($pub(id_i, R)$), which guarantees that the new replica will receive the resources. When the synchronization procedure has ended, the replica will be automatically kept synchronized with the IoT Hub during the normal system lifecycle.

When resources are polled for (as will be described in Section 6.6.1.2), the IoT Hub might find out that a resource targeted by some request has changed. A request targeting a resource that either has not been cached or is not considered fresh must be forwarded by the IoT Hub to the smart object. Upon receiving the response from the smart object, after updating its cache and forwarding the response to the requesting replica, the IoT Hub uses the synchronization protocol to publish the updated information to all of its replicas.

When observing resources (as described in Section 6.6.1.2), the synchronization procedure resembles that for the polling case. When resources are pushed by smart objects to the IoT Hub (as described in Section 6.6.1.2), the IoT Hub uses the synchronization protocol to publish the updated information at all replicas. Note that this synchronization strategy is needed only for those replicas that are part of the request/response loop. In fact, as all resources are automatically synchronized with the request-issuing replica by design, the replica perfectly reproduces the behavior of the actual IoT Hub.

In order to validate the feasibility of the proposed IoT architectural solution and to evaluate its performance, extensive experimentation was conducted. The evaluation focused on the resource management on both local and remote IoT Hubs and the synchronization mechanisms previously described

The experimental setup was designed and deployed with the aim of creating a realistic scenario, with heterogeneous components and nodes in a local IoT network and the cloud. The main components are:

- *Smart objects*: Real and virtual nodes with CoAP modules based on the Californium framework [94].
- *IoT Hubs*: Raspberry Pi model B [223] or independent VM instance running all the functional modules presented in Section 6.6.1.1 (resource discovery, proxy CoAP/CoAP and HTTP/CoAP, border router, cache and replica manager).
- *Virtualization platforms*: Four different virtualization configurations on both local and cloud platforms are considered:
  – Microsoft Azure [224]
  – Amazon EC2 [225]

- Open Stack [226] on Microsoft Azure
- Open Stack on a local physical machine.
- *Resource external consumer*: Real and virtual external consumers implementing HTTP and CoAP modules to dynamically interact and consume resources managed by the platform and active IoT Hubs and smart objects.

We configured and tested multiple virtualization configurations in order to evaluate the performance of the designed IoT architecture both on local and remote VMs. In particular, the Open Stack layer was tested on a local installation at the Department of Information Engineering of the University of Parma and, at a later stage, on Microsoft Azure in order to obtain and measure more realistic results on a professional cloud infrastructure. The local Open Stack installation runs on a physical machine with two 1.6 GHz processors and 3 GB RAM, while the Azure configuration was a VM with four 2.0 GHz cores and 8 GB RAM.

These two platforms were used to dynamically manage replicas of active IoT Hubs and to handle resource synchronization and remote data access. A virtual instance of an IoT Hub replica is characterized by an hardware profile with a single-core 2 GHz processor; 1 GB RAM; and 8 GB disk space. The internal IoT Hub runs a Linux Ubuntu 14.04 LTS operating system with SSH remote access, Oracle Java VM [227], and all the required functional software modules already installed and configured.

The following key metrics are defined to measure the performance at different architectural layers:

- *IoT Hub replica creation time*: the time required to create and run, on the target cloud infrastructure, a new instance of an IoT Hub replica.
- *Resource synchronization time*: the elapsed time needed to synchronize a new resource between two IoT Hubs.
- *Resource access time*: the time required to access and retrieve a response for a resource of interest. It can be associated with different configurations:
  - direct and local access to the CoAP server smart object (e.g., if the consumer and the node are in the same network);
  - remote access through the cloud and communication with the physical hub;
  - remote access to the cached value stored on an IoT Hub replica.

- *CPU usage %*: the percentage of CPU used by the IoT Hub core process.
- *Memory usage %*: the memory percentage used by the core process of the IoT Hub.

The first phase of the experimental performance analysis focuses on the evaluation of the time required to create (from scratch) and run a new IoT Hub replica instance on different virtualized cloud infrastructures. The results are shown Figure 6.30. Each value has been obtained by averaging over 10 different VM creation runs and has a confidence interval of 99%. It can be observed that:

- the average costs on remote and professional cloud infrastructures are comparable;
- the cost is higher on local and non-optimized solutions (such as the Open Stack instance running in our department).

We note that the metric corresponds to the total amount of time required to create a new VM from scratch (starting from a pre-configured image and adding the time to start all the required services and architectural software processes). This cost should be considered only once for each IoT Hub replica and is significantly written off with the increase of the hub's lifetime. Native VMs on Microsoft Azure and Amazon EC2 have approximately the same creation time, while the Open Stack platform introduces a small delay associated with the additional overhead (of the platform itself) required to manage multiple servers.

The second phase of the experimental analysis was focused on:

- the evaluation of the time required for synchronization of resources between two IoT hbs (physical and virtual);
- the time needed by an external consumer to access a target resource of interest in different scenarios and configurations.

Figure 6.29a shows, as functions of the number resources:

- the total time required for the synchronization of a set of resources between an IoT Hub and its new replica;
- the average time required to synchronize a single resource in the same set.

The results show that the average synchronization cost for a single resource is stable and, consequently, the total required time is proportional to the number of resources to be synchronized. The results

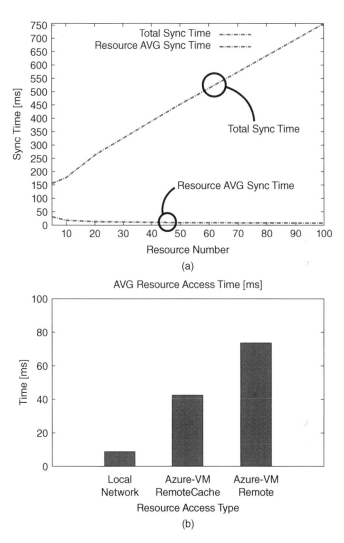

Figure 6.29 (a) Average synchronization time with respect to the number of synchronized resources; (b) average remote resource access time in different application scenarios.

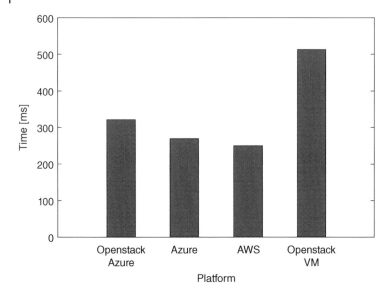

Figure 6.30 Average IoT Hub creation time on different cloud platforms.

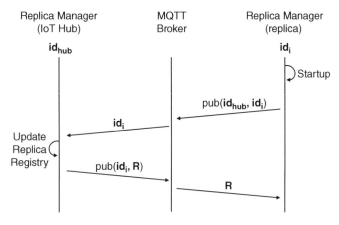

Figure 6.31 Synchronization procedure performed at startup of the replica of an IoT Hub.

also show that for a reduced number of resources, the impact of the cost of MQTT connection creation is only slightly relevant compared with the payload. Obviously, by increasing the number of resources synchronized using the same MQTT connection it is possible to reduce this effect and reach a stable value below 25 ms/resource.

In Figure 6.29b, the average time required by a consumer to access a resource of interest provided by a smart object is evaluated in different scenarios. In particular, we have considered three different configurations where:

- the consumer is in the same local network as the target smart object;
- the external actor accesses a cached value on the active IoT Hub replica on the Azure platform;
- the consumer accesses a resource that is not cached on the IoT Hub replica and, consequently, there is a requirement for direct communication between the virtual replica and the real IoT Hub.

The results show, as intuitively expected, that the quickest access is obtained if the consumer is in the same network as the hub and does not require additional communication with remote infrastructure. However, if the consumer is an external actor, the average cost – still considering a realistic deployment through Microsoft Azure – is below 80 ms/resource. This value decreases if we consider a resource that is cached on the replica hub and does not require any additional communication with the local hub and, eventually, the smart object.

Our experimentation also investigated the cost, in terms of CPU usage of an IoT Hub process, both for local (Raspberry Pi node) and remote (Microsoft Azure VM) instances. Figure 6.32 show the percentage of CPU usage during a 60-s run of core activities, highlighting:

- initialization of the main process and the creation of a set of five new resources on the local hub;
- activation of the remote replica and its initialization;
- synchronization of the group of five initial resources between local and remote replica hub;
- sporadic addiction of single resources until the end of the experiment.

The results show how the initialization of the hub is an intensive activity on both local and remote instances. The CPU usage incurs a significant one-time cost due to:

- setting up the hub configurations (such as node identification and software module selection);
- establishing the VPN connection activating the CoAP server and the MQTT module (listener and publisher on specific topics);
- starting up the resource directory.

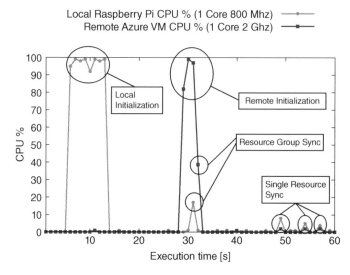

Figure 6.32 IoT Hub process CPU percentage usage for local (Raspberry Pi node) and remote (Microsoft Azure VM) instances.

After the initialization phase, the CPU usage significantly reduces for both resource group synchronization and sporadic management of new single resources. These activities represent common and frequent tasks for an active IoT Hub, which typically handles small variations in the set of managed resources during its lifetime. The main software process consumes a reduced amount of CPU (under 10%) for a small amount of time on both local and remote instances.

In order to complete the analysis, Table 6.3 shows the average values (obtained through multiple independent runs and with a confidence

Table 6.3 Average CPU and memory utilization related to specific IoT Hub procedures on both local and remote instances.

Hub	Init %	Resource add %	Sync resource group %	Sync single resource %
[Local] Raspberry Pi CPU	96	17.45	15.15	6.93
[Local] Raspberry Pi Memory	+2,7	+0.02	+0.01	+0.002
[Remote] Azure VM Remote CPU	97	NA	34.22	1.93
[Remote] Azure VM Memory	+1.9	NA	+0.01	+0.001

interval of 99%) of CPU and memory usage related to each specific hub procedure. The data confirm the cost distribution, with a percentage peak due to the initialization phase and lower values for group and single-resource synchronization. Memory utilization has been measured as the offset with respect to the previous value and depends on the Java VM memory management [228]. In particular, when an object is no longer used, the Java Garbage Collector reclaims the underlying memory and reuses it for future object allocation without explicit deletion (no memory is given back to the operating system).

An important aspect of a truly scalable architecture is the ability to quickly react to dynamic load changes, characterized by the rate of incoming requests that need to be served. The replication strategy proposed in this paper aims at providing a flexible and efficient management of IoT Hub replicas in order to guarantee that the response time remains below a given threshold. The virtualization approach provides a flexible solution that guarantees significant availability and efficient load balancing in highly dynamic IoT environments. In order to validate the designed replica management scheme, an additional experimental phase was carried out. Leveraging the same Microsoft Azure infrastructure used for all the other experiments, we measure the response time for incoming requests for smart object resources managed by the IoT Hub.

In Figure 6.33, the effect of replica management is shown in terms of response time as a function of the incoming request rate. In particular, the results are the smallest number of requests that prompts creation of a new replica of the IoT Hub, which happens whenever the response time exceeds a threshold of 800 ms. The graph clearly shows that the response time tends to increase linearly as a function of the rate of requests until the IoT Hub reaches a breaking point associated with the maximum number of requests it can handle. This is clearly visible in the areas of the graph characterized by steep slopes. When a slope is detected, we activate a new replica, which brings the response time back to a value that meets our operational requirements. As the request rate increases, new replicas are created. As shown in Figure 6.33, we stopped our experimentation after the activation of three replicas. It is worth noting that:

- a new replica is activated almost periodically (every 800 ms, according to the set threshold)
- the slope of the response time between two consecutive activations is inversely proportional to the number of active replicas.

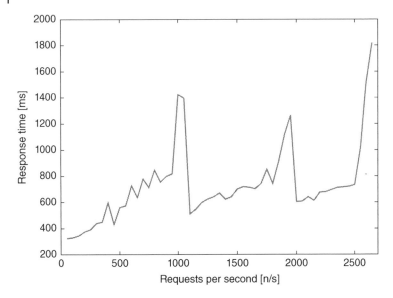

Figure 6.33 Effect of replica management with respect to the increasing number of requests per second (Microsoft Azure infrastructure).

The widespread adoption of container-based technologies has significantly changed the way cloud and fog applications can be deployed and delivered to final users. In order to provide a thorough performance evaluation of the proposed solution, we also created a container-based version of the IoT Hub using the Docker platform [216]. The container has the same networking configuration, features, and services as running on the VM-based version but shares an operating system kernel and features with other container instances running on the same server. The local experimentation was conducted using a VM running Ubuntu 14.04, with one 2.0 GHz processor and 1 GB RAM. Docker 1.11 was executed on top of this with the aim of evaluating the startup time of the dockerized IoT Hub. Such a low-end hardware profile, compared to realistic data center facilities, was purposely chosen to show the small footprint of the IoT Hub.

In Figure 6.34, the average total startup time required to activate a fully operative IoT Hub instance, as well as the breakdown of container and IoT Hub services startup time, are shown. The results have been averaged over 1000 runs on the configured setup. The container startup time simply considers the activation of a Docker

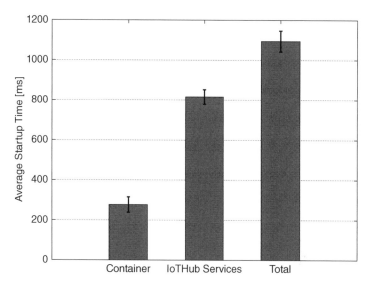

Figure 6.34 Average IoT Hub startup time on a Docker container.

container instance. On top of this running instance, we measure the time required to activate all IoT Hub services and to respond to incoming HTTP and CoAP requests. The results show how a container-based approach can be efficiently adopted both on cloud or fog infrastructures (according to the target application scenarios) to support efficient and highly dynamic creation and management of IoT Hub replicas on top of existing host machines. Unlike a VM-based approach, container-based IoT Hub instances can be instantiated and removed dynamically, depending on the instantaneous load and without affecting the host machine or requiring infrastructure-related efforts (such as file/template management or configuration and activation of new machines).

7

The IoT in Practice

7.1   Hardware for the IoT

The IoT is expected be a worldwide network comprising, by 2020, billions of devices. This gigantic number of devices, pervasively deployed, will be characterized by their heterogeneity in terms of software and, in particular, hardware.

In order to provide a general definition for hardware platforms, Figure 7.1 shows an high-level view of the main hardware components in a smart object. The illustrated modules are:

- *Communication module*: This gives the smart object its communication capabilities. It is typically either a radio transceiver with an antenna or a wired connection.
- *Microcontroller*: This gives the smart object its behavior. It is a small microprocessor that runs the software of the smart object.
- *Sensors or actuators*: These give the smart object a way to sense and interact with the physical world.
- *Power source*: This is needed because the smart object contains electrical circuits. The most common power source is a battery, but there are other examples as well, such as piezoelectric power sources, that provide power when a physical force is applied, or small solar cells that provide power when light shines on them.

Microcontrollers have two types of memory: Read-only memory (ROM) and random access memory (RAM). ROM is used to store the program code that encodes the behavior of the device and RAM is used for temporary data the software needs to do its task. For

Internet of Things: Architectures, Protocols and Standards, First Edition.
Simone Cirani, Gianluigi Ferrari, Marco Picone, and Luca Veltri.

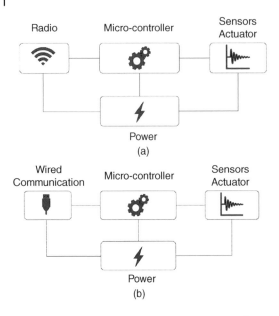

Figure 7.1 Smart object hardware, with (a) radio network interface and (b) wired communication interface.

example, temporary data includes storage for program variables and buffer memory for handling radio traffic.

For constrained devices, the content of the ROM is typically burned into the device when it is manufactured and is not altered after deployment. Modern microcontrollers provide a mechanism for rewriting the ROM, which is useful for in-field updates of software after the devices have been deployed.

In addition to memory for storing program code and temporary variables, microcontrollers contain a set of timers and mechanisms for interacting with external devices, such as communication devices, sensors, and actuators. The timers can be freely used by the software running on the microcontroller. External devices are physically connected to its pins. The software communicates with the devices using mechanisms provided by the microcontroller, typically in the form of a serial port or a serial bus. Most microcontrollers provide a so-called universal synchronous/asynchronous receiver/transmitter (USART) for communication with serial ports. Some USARTs can be configured to work as a serial peripheral interface (SPI) bus for communicating with sensors and actuators.

A smart object is driven by electronics, and electronics need power. Therefore, every smart object needs a power source (some power

Table 7.1 Power sources for smart objects, maximum current draws, and charge they can store.

Power source	Typical maximum current (mA)	Typical charge (mAh)
CR2032 button cell	20	200
AA alkaline battery	20	3000
Solar cell	40	Limitless
RF power	25	Limitless

source examples are reported in Table 7.1). Today, the most common power source is a battery, but there are several other possibilities for power, such as solar cells, piezoelectricity, radio-transmitted energy, and other forms of power scavenging. Lithium cell batteries are currently the most common. With low-power hardware and proper energy-management software, a smart object can have a lifetime of years on standard lithium cell batteries. Unlike cell phones and laptops, which are human-operated, most smart objects are designed to operate without human control or human supervision. Furthermore, many smart objects are located in difficult-to-reach places, and many are embedded in other objects. Therefore, in most cases it is impractical to recharge their batteries.

7.1.1 Classes of Constrained Devices

Despite the overwhelming variety of Internet-connected devices envisioned, it is absolutely worthwhile to have a common terminology for different classes of constrained devices. For this reason, the RFC7228 [229] provides a definition for the main classes and characteristics of IoT smart objects and constrained devices (Table 7.2).

These characteristics correspond to distinguishable clusters of commercially available chips and design cores for constrained devices. It is expected that the boundaries of these classes will move over time; Moore's law tends to be less effective in the embedded space than in personal computing devices, so gains made available by increases in transistor count and density are more likely to be invested in reductions in cost and power requirements than in continual increases in computing power.

Table 7.2 RFC7228 classes of constrained devices.

Name	Data size (e.g., RAM)	Code size (e.g., Flash)
Class 0 (C0)	≪10 KiB	≪100 KiB
Class 1 (C1)	~ 10 KiB	~ 100 KiB
Class 2 (C2)	~ 50 KiB	~ 250 KiB

In more detail, the classes are as follows:

Class 0

devices are very constrained sensor-like motes. They are so severely constrained in memory and processing capabilities that most likely they will not have the resources required to communicate directly with the Internet in a secure manner. These devices will participate in Internet communications with the help of larger devices acting as proxies, gateways, or servers. Class 0 devices generally cannot be secured or managed comprehensively in the traditional sense. They will most likely be preconfigured (and will be reconfigured rarely, if at all) with a very small data set. For management purposes, they could answer keep-alive signals and send on/off or basic health and status indications.

Class 1

devices are quite constrained in code space and processing capabilities, such that they cannot easily talk to other Internet nodes employing a full protocol stack such as HTTP, Transport Layer Security, and related security protocols and XML-based data representations. They are can use a protocol stack specifically designed for constrained nodes (such as CoAP over UDP) and participate in meaningful conversations without the help of a gateway node. They can provide support for the security functions required on a large network. Therefore, they can be integrated as fully developed peers into an IP network, but they need to be parsimonious with state memory, code space, and often power expenditure for protocol and application usage.

Class 2

devices are less constrained and fundamentally capable of supporting most of the same protocol stacks as used on notebooks or servers. They can benefit from lightweight and energy-efficient protocols and from consuming less bandwidth. Using fewer resources for networking leaves more resources available for applications. Thus, using the protocol stacks defined for more constrained devices on Class 2 devices might reduce development costs and increase interoperability.

7.1.2 Hardware Platforms

In this section we introduce some of the main hardware platforms available on the market, trying to highlight their distinctive features and associated classes.

7.1.2.1 TelosB

TelosB[1] is a mote from Memsic Technology. It has the same design as the Tmote Sky mote from Sentilla. It is comprises the MSP430 (the MSP430F1611) microcontroller and a CC2420 radio chip. The microcontroller of this mote operates at 415 MHz and has 10 kB internal RAM and a 48 kB programmable flash memory.

The TelosB was developed by the University of California, Berkeley. It was a new mote design based on experiences with previous mote generations. It was designed with three major goals that would enable experimentation: minimal power consumption, ease of use, and increased software and hardware robustness. The use of the MSP430 in Telos gave it a power profile almost one-tenth that of previous mote platforms.

Figure 7.2 is a schematic overview of the mote and shows how the components interact. Table 7.3 gives the detailed hardware profile, with the modules and their associated descriptions. Telos B can be classified as a Class 0 constrained device.

7.1.2.2 Zolertia Z1

The Zolertia Z1[2] is a general purpose development board targeting wireless sensor networks and heterogeneous IoT applications. It is equipped with two on-board digital sensors (an accelerometer and a temperature sensor), and uses Phidget Sensors connectors to easily extend connected devices such as sensors and actuators.

Figure 7.3 and Table 7.4 are a schematic overview of the Z1, and a summary of the components interactions and available sensors with

---

1 http://www.memsic.com/userfiles/files/Datasheets/WSN/
telosb:datasheet.pdf.
2 http://zolertia.io/z1.

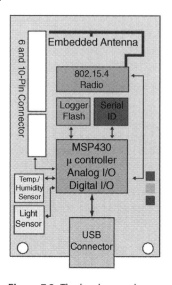

Figure 7.2 The hardware schema and the board image of the TelosB mote platform.

Table 7.3 Hardware specification of the MemSic TelosB mote.

Name	Description
MCU	TI MSP430F1611
RAM	10 kB
ROM	48 kB
Serial communication	UART
Main module current draw	1.8 mA (active mode)
	5.1 µA (sleep mode)
IEEE 802.15.4 compliant	
RF transceiver	2400–2483.5 MHz
RF current draw	23 mA (receive mode)
	21 µA (idle mode)
	1 µA (sleep mode)
Battery	2 × AA batteries
Sensors	Visible light sensor
	Humidity sensor
	Temperature sensor

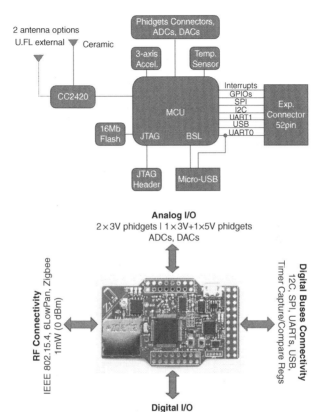

Analog I/O
2 × 3V phidgets | 1 × 3V+1×5V phidgets
ADCs, DACs

RF Connectivity
IEEE 802.15.4, 6LowPan, Zigbee
1mW (0 dBm)

Digital Buses Connectivity
I2C, SPI, UARTs, USB,
Timer Capture/Compare Regs

Digital I/O
GPIOs, Interrupts, Timers, Comparators I/O
Figure 3 — *Z*1 *Expansion Capabilities (54-pin XPCon)*

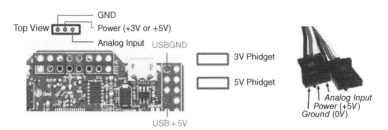

Figure 7.3 The Zolertia Z1 platform with board images and main component schema.

Table 7.4 Hardware specification of the Zolertia Z1 platform.

Name	Description
MCU	TI MSP430F2617
RAM	8 kB
ROM	92 kB
Digital communication	I2C, SPI and UART
Main module current draw	0.5 mA (active mode)
	0.5 μA (standby mode)
IEEE 802.15.4 compliant	
RF transceiver	CC2420 2.4 GHz
RF current draw	18.8 mA (receive mode)
	426 μA (idle mode)
	20 μA (sleep mode)
Battery	2 × AA or AAA cells
	1 × CR2032 coin cell
Sensors	Low-power digital temperature sensor 3-axis, ±2/4/8/16*g* digital accelerometer, 3 V and 5 V Phidget Sensors connectors

their connectors. Like Telos B, Zolertia Z1 can be classified as a Class 0 constrained device.

7.1.2.3 OpenMote

OpenMote[3] hardware is composed of three boards: OpenMote-CC2538, OpenBase and OpenBattery. OpenMote-CC2538 is the mote itself and includes the microcontroller and the radio transceiver, as well as other peripherals such as LEDs and buttons. The OpenBase is the board allowing programming and debugging through a UART or USB interface with a computer or via an Ethernet port with the Internet. The OpenBattery is the board that lets OpenMote-CC2538 run autonomously by providing energy to all its subsystems, as well allowing it to interface it with various sensors.

---

3 http://www.openmote.com.

The OpenMote-CC2538 includes the following hardware:

- *CC2538*: This is a system on a chip (SoC) from Texas Instruments, with a 32-bit Cortex-M3 microcontroller and a CC2520-like radio transceiver. The microcontroller runs up to 32 MHz and includes 32 kB of RAM and 512 kB of flash memory, and the usual peripherals (GPIOs, ADC, timers, etc.). The radio operates in the 2.4 GHz band and is fully compatible with the IEEE 802.15.4-2006 standard.
- *TPS62730*: This is a step-down DC/DC converter from Texas Instruments with two operation modes: regulated and bypass. In bypass mode the TPS62730 directly connects the input voltage from the battery (typically 3 V) to the whole system. In regulated mode the TPS62730 regulates the input voltage down to 2.1 V. The benefit of this architecture is in terms of system efficiency, since it is an improvement under both low- and high-load conditions; that is, either when the system is sleeping or when the radio is transmitting or receiving.
- *ABM8G*: This is a 32 MHz crystal from Abracon Corporation, used to clock the microcontroller and the radio transceiver. It is rated at 30 ppm (parts per million) from −20°C to +70°C.
- *ABS07*: This is a 32.768 kHz crystal from Abracon Corporation used to clock the microcontroller's real time clock. It is rated at 10 ppm from −40°C to +85°C.
- *LEDs*: There are four LEDs (red, green, yellow and orange) from Rohm Semiconductor, used for debugging purposes.
- *Buttons*: There are two buttons, from Omron. One is used to reset the board and the other is connected to a GPIO line, thus enabling the microcontroller to be woken from sleep modes through an interrupt.
- *Antenna connector*: The antenna connector enables an external antenna to be connected to the board.
- *XBee layout*: The OpenMote is fully compliant with the XBee form factor, meaning that it can be easily interfaced with a computer using a XBee Explorer dongle.

Figure 7.4 and Table 7.5 give a schematic overview of the Open Mote, the component interactions and the available sensors with connectors. The OpenMote runs multiple operating systems, such as Contiki, OpenWSN, FreeRTOS and RiOT. It can be classified as a Class 0 constrained devices.

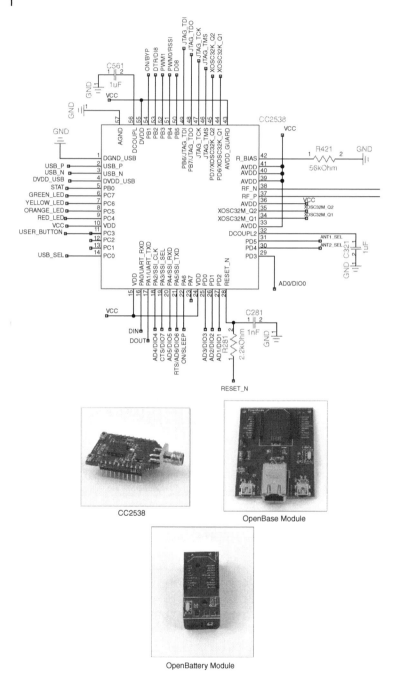

Figure 7.4 The OpenMote platform with main hardware schema and core board with battery and OpenBase additional modules.

Table 7.5 Hardware specification of the OpenMote platform.

Name	Description
MCU	TI 32-bit Cortex-M3
RAM	32 kB
ROM	512 kB
Digital communication	I2C, and UART
Main module current draw	0.5 mA (active mode)
	0.5 μA (standby mode)
IEEE 802.15.4 compliant	
RF transceiver	CC2520 2.4 GHz
Battery	2 × AAA cells
Sensors	Temperature/humidity sensor (SHT21)
	Acceleration sensor (ADXL346)
	Light sensor (MAX44009)

7.1.2.4 Arduino

Arduino[4] is a computer hardware and software company manufacturing microcontroller kits for building digital devices that can sense and interact with the physical world. The board designs adopt a variety of microprocessors and controllers and are equipped with sets of digital and analog input/output pins, which may be interfaced expansion boards (called "shields") and other external circuits and components. The typical programming language is a dialect of the traditional C and C++, with the possibility of including the many existing libraries from the developer community.

The Arduino project started in Italy in 2005 as a program for students at the Ivrea Interaction Design Institute. The aim was to create a low-cost and easy-to-use board for novices and professionals. Arduino boards have been designed to create devices and prototypes that interact with the environment using sensors and actuators and multiple communication paradigms (thanks to the shield expansion-board system).

One of the first Arduino boards built for the IoT ecosystem was the Arduino Yun. This is a microcontroller board based on the

---

4 https://www.arduino.cc.

ATmega32u4 and the Atheros AR9331. The Atheros processor runs a Linux distribution, based on OpenWrt,[5] called Linino OS. The board has built-in Ethernet and Wi-Fi communication interfaces, a USB-A port, a micro-SD card slot, 20 digital input/output pins (seven of which can be used as PWM outputs and twelve as analog inputs), a 16 MHz crystal oscillator, a micro USB connection, an ICSP header, and three reset buttons.

Figure 7.5 shows the Arduino Yun architecture and how the Linux module of the board can communicate with a traditional Arduino module. This communication capability distinguishes this board from the other boards, offering a powerful networked computer that can be used to create IoT prototypes in several application scenarios.

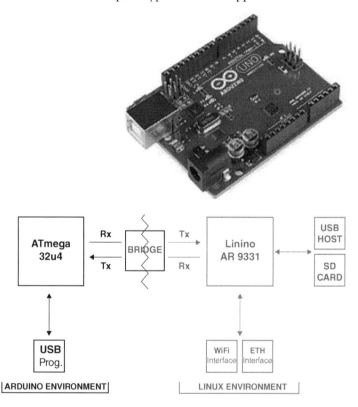

Figure 7.5 Classical and Yun versions of the Arduino platform with a detailed representation of the Yun hardware architecture and main components.

---

5 https://openwrt.org/.

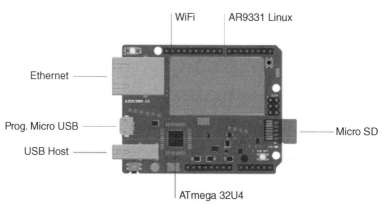

Figure 7.5 (Continued)

Tables 7.6 and 7.7 report the detailed hardware profile of Arduino Yun modules with their associated components and descriptions. The Yun board can be classified as a Class 2 constrained device.

7.1.2.5 Intel Galileo

Intel Galileo[6] was the first Arduino-certified development board based on Intel's x86 architecture. It was designed for makers and education/

6 https://software.intel.com/en-us/iot/hardware/galileo.

Table 7.6 Hardware specification of the Arduino Yun AVR Arduino microcontroller.

Name	Description
Microcontroller	ATmega32U4
Operating Voltage	5 V
Input Voltage	5 V
Digital I/O pins	20
PWM Output	7
Analog I/O pins	12
Flash memory	32 kB (4 kB used by the bootloader)
SRAM	2.5 kB
EEPROM	1 kB
Clock speed	16 MHz

Table 7.7 Hardware specification of the Arduino Yun Arduino microprocessor.

Name	Description
Processor	Atheros AR9331
Architecture	MIPS
Operating voltage	3.3 V
Ethernet	IEEE 802.3 (10/100Mbit/s)
Wi-Fi	IEEE 802.11b/g/n (2.4 GHz)
USB type	2.0 Host
Flash memory	16 MB
RAM	64 MB DDR2
SRAM	2.5 kB
EEPROM	1 kB
Clock speed	400 MHz

academic communities. The board combines Intel technology with support for Arduino expansion shields and the related software and libraries. The board runs an open-source Linux operating system with the Arduino software libraries, enabling re-use of existing software. Intel Galileo hosts a Linux operating system.

Intel Galileo is equipped with the Intel Quark SoC X1000, the first product from the Intel Quark technology family of low-power, small-core products. The Galileo board comes with several computing industry standard I/O interfaces, including ACPI, PCI Express, 10/100 Mbit Ethernet, Micro SD or SDHD, USB 2.0 and EHCI/OHCI USB host ports, high-speed UART, RS-232 serial port, programmable 8 MB NOR flash, and a JTAG port for debugging.

Figure 7.6 and Table 7.8 give a schematic overview of the Intel Galileo boards, component interactions and available communication modules, for both first- and second-generation devices. Galileo can be classified as a Class 2 constrained devices.

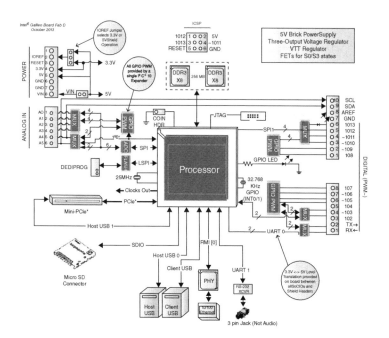

Figure 7.6 First and second generations of the Intel Galileo platform with board schemas and hardware architecture of main components.

Gen 1

Gen 2

Figure 7.6 (*Continued*)

7.1.2.6   Raspberry Pi

The Raspberry Pi[7] is a series of single-board computers created in the United Kingdom by the Raspberry Pi Foundation. The original aim of

_____

7 https://www.raspberrypi.org/.

Table 7.8 Hardware specification of the Intel Galileo board.

Name	Description
MCU	SoC X Intel Qua k X1000
RAM	256 MB
Memory	SDCard (MBytes)
Serial communication	I2C and UART
Main module current draw	0.5 mA (active mode)
	0.5 μA (standby mode)
Network adapter	IEEE 802.3 10/100 (Ethernet)
Battery	–
Sensors	–

the founders was to encourage the teaching of basic computer science in schools and developing countries. Step by step, their boards significantly changed the way manufacturers and developers thought about and created new projects in many application scenarios. For example, the original model became very popular and started spreading outside of its initial target market for uses such as robotics.

Figure 7.7 shows three Raspberry Pi boards with their available components and hardware profiles. The Broadcom BCM2835 SoC was used on the first generation and was inspired by the chip used in first-generation smartphones (its CPU is an older ARMv6 architecture). It includes a 700 MHz ARM1176JZF-S processor, VideoCore IV GPU, and RAM. It has a level 1 (L1) cache of 16 kB and a level 2 (L2) cache of 128 kB. The level 2 cache is used primarily by the GPU. The SoC is stacked underneath the RAM chip, so only its edge is visible.

The Raspberry Pi 2 uses a Broadcom BCM2836 SoC with a 900 MHz 32-bit quad-core ARM Cortex-A7 processor (as do many current smartphones), with 256 kB shared L2 cache.

The Raspberry Pi 3 uses a Broadcom BCM2837 SoC with a 1.2 GHz 64-bit quad-core ARM Cortex-A53 processor, with 512 kB shared L2 cache.

The Model A, A+ and Pi Zero are shipped without Ethernet modules and are commonly connected to a network using external adapters for Ethernet or Wi-Fi. Models B and B+ have the Ethernet port that is provided by a built-in USB Ethernet adapter using the

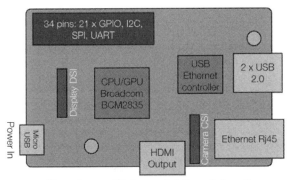

Raspberry Pi 1 model B revision 2

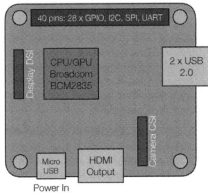

Raspberry Pi 1 model A+ revision 1.1

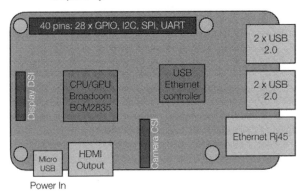

Raspberry Pi 1 model B+ revision 1.2
and Raspberry Pi 2 model B

Figure 7.7 Main Raspberry Pi boards and revisions.

SMSC LAN9514 chip. The Raspberry Pi 3 and Pi Zero W (wireless) provide a 2.4 GHz Wi-Fi 802.11n (150 Mbit/s) and Bluetooth 4.1 (24 Mbit/s) connectivity module based on a Broadcom BCM43438 chip. The Raspberry Pi 3 is also equipped with a 10/100 Ethernet port.

The Raspberry Pi can be also used with USB storage, USB-to-MIDI converters, and virtually any other device/component with USB capabilities. Other external devices, sensors/actuators and peripherals can be attached through a set of pins and connectors available on the board's surface.

The Raspberry Pi board family can run multiple operating systems, such as Raspbian, Fedora, Ubuntu MATE, Kali Linux, Ubuntu Core, Windows 10 IoT Core, RISC OS, Slackware, Debian, Arch Linux ARM, and Android Things. This combination of high-profile hardware, software and operating systems makes these boards to represent powerful and complex nodes in heterogenous IoT applications. They can efficiently work as IoT Hubs, gateways and data collectors using heterogenous protocols and running multiple services at the same time. All the Raspberry Pi boards can be classified as Class 2 constrained devices.

7.2   Software for the IoT

In this section, an overview of the main operating systems for the IoT is presented. The Contiki operating system is particularly important, and a more detailed description is therefore provided in Section 7.2.6.

7.2.1   OpenWSN

The OpenWSN[8] project is an open-source implementation of a fully standards-based protocol stack for IoT networks. It was based on the new IEEE802.15.4e time-slotted channel-hopping standard. IEEE802.15.4e, coupled with IoT standards such as 6LoWPAN, RPL and CoAP, enables ultra-low-power and highly reliable mesh networks that are fully integrated into the Internet.

OpenWSN has been ported to numerous commercial available platforms from older 16-bit micro-controllers to state-of-the-art 32-bit

---

8 https://openwsn.atlassian.net/wiki/pages/viewpage.action?
pageId=688187.

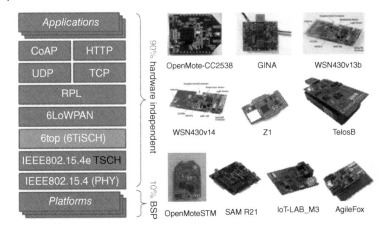

Figure 7.8 OpenWSN protocol stack, highlighting hardware-independent modules and supported hardware platforms.

Cortex-M architectures. The OpenWSN project offers a free and open-source implementation of a protocol stack and the surrounding debugging and integration tools, thereby contributing to the overall goal of promoting the use of low-power wireless mesh networks.

Figure 7.8 shows the OpenWSN protocol layers and software libraries, which are hardware independent, and a set of supported hardware platforms, on which it can be installed and used.

7.2.2 TinyOS

TinyOS[9] is a free, open-source, BSD-licensed OS designed for low-power embedded distributed wireless devices used in sensor networks. It has designed to support the intensive concurrent operations required by networked sensors, with minimal hardware requirements. TinyOS was developed by University of California, Berkeley, Intel Research, and Crossbow Technology. It is written in the nesC (Network Embedded Systems C) programming language, which is a version of C optimized to support components and concurrency. It is also component-based, supporting event-driven programming of applications for TinyOS.

---

9 www.tinyos.net/.

7.2.3 FreeRTOS

FreeRTOS[10] is a real-time operating system kernel for embedded devices designed to be small and simple. It been ported to 35 micro-controllers and it is distributed under the GPL with an optional exception. The exception permits users' proprietary code to remain closed source while maintaining the kernel itself as open source, thereby facilitating the use of FreeRTOS in proprietary applications.

In order to make the code readable, easy to port, and maintainable, it is written mostly in C, (but some assembly functions have been included to support architecture-specific scheduler routines). It provides methods for multiple threads or tasks, mutexes, semaphores and software timers.

7.2.4 TI-RTOS

TI-RTOS[11] is a real-time operating system that enables faster development by eliminating the need for developers to write and maintain system software such as schedulers, protocol stacks, power-management frameworks and drivers. It is provided with full C source code and requires no up-front or runtime license fees. TI-RTOS scales from a low-footprint, real-time preemptive multitasking kernel to a complete RTOS with additional middleware components including a power manager, TCP/IP and USB stacks, a FAT file system, and device drivers, allowing developers to focus on differentiating their applications.

Figure 7.9 shows the main software components of the TI-RTOS operating system. In particular, it is based on a core layer with a real-time kernel, connectivity support and power management. On top of that, a set of platform APIs allow the developer to build custom applications. The OS provides a large set of ready-to-use libraries based on TCP/UDP/IP networking, standard BSD socket interface and main application layer protocols such as HTTP, TFTP, Telnet, DNS, and DHCP.

---

10 www.freertos.org/.
11 http://www.ti.com/tool/ti-rtos.

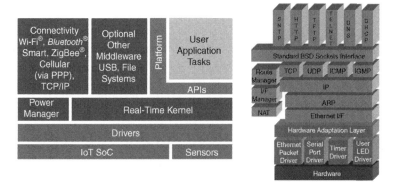

Figure 7.9 Schematic overview of the TI-RTOS operating system with main modules and software components.

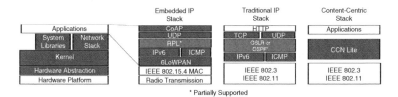

Figure 7.10 Overview of networking architecture for the RIOT operating system.

7.2.5 RIOT

RIOT[12] is an open-source microkernel operating system for the IoT, licensed as LGPL. It allows C and C++ application programming, and provides both full multi-threading and real-time capabilities (in contrast to other operating systems with similar memory footprints, such as TinyOS or Contiki). RIOT runs on 8-bit (e.g., AVR Atmega), 16-bit (e.g., TI MSP430) and 32-bit hardware (e.g., ARM Cortex). A native port also enables RIOT to run as a Linux or MacOS process, enabling the use of standard development and debugging tools such as GNU Compiler Collection, GNU Debugger, Valgrind, Wireshark, an so on. RIOT is partly POSIX-compliant and provides multiple network stacks, including IPv6, 6LoWPAN and standard protocols such as RPL, UDP, TCP, and CoAP (see Figure 7.10).

---

12 www.riot-os.org/.

7.2.6   Contiki OS

Contiki is an operating system for networked, memory-constrained systems, targeting low-power wireless IoT devices. Its main characteristics are:

- It is open source and in continuous development. Even if it is less well documented and less well maintained than commercial operating systems, it allows developers not only to work on custom applications but also to modify core OS functionalities such as the TCP/IP stack and the routing protocol.
- It provides a full TCP/uIPv6 stack using 6LoWPAN [230] for header compression, and creates LR-WPAN routes with RPL [39], the IPv6 routing protocol for low-power and lossy networks

Contiki was created by Adam Dunkels in 2002 and has been further developed by a worldwide team of developers from Texas Instruments, Atmel, Cisco, ENEA, ETH Zurich, Redwire, RWTH Aachen University, Oxford University, SAP, Sensinode, the Swedish Institute of Computer Science, ST Microelectronics, Zolertia, and many others.

Contiki is designed to run on classes of hardware devices that are severely constrained in terms of memory, power, processing power, and communication bandwidth. For example, in terms of memory, despite providing multitasking and a built-in TCP/IP stack, Contiki only needs about 10 kB of RAM and 30 kB of ROM. A typical Contiki system has memory of the order of kilobytes, a power budget of the order of milliwatts, processing speed measured in megahertz, and communication bandwidth of the order of hundreds of kilobits/second. This class of systems includes various types of embedded systems as well as a number of old 8-bit computers.

A brief description of the core features of Contiki will be provided, highlighting why they are of particular interest for building complex IoT applications.

7.2.6.1   Networking

Contiki provides three network mechanisms:

- the uIP[13] TCP/IP stack, which provides IPv4 networking;
- the uIPv6 stack, which provides IPv6 networking;
- the Rime stack, which is a set of custom lightweight networking protocols designed specifically for low-power wireless networks.

---

13 https://github.com/adamdunkels/uip.

The IPv6 stack was contributed by Cisco and was, at the time of release, the smallest IPv6 stack to receive IPv6-ready certification. The IPv6 stack also contains the RPL routing protocol and the 6LoWPAN header compression and adaptation layer.

The Rime stack is an alternative network stack that is intended to be used when the overhead of the IPv4 or IPv6 stacks is prohibitive. The Rime stack provides a set of communication primitives for low-power wireless systems. The default primitives are single-hop unicast, single-hop broadcast, multi-hop unicast, network flooding, and address-free data collection. The primitives can be used on their own or combined to form more complex protocols and mechanisms.

7.2.6.2   Low-power Operation

Many Contiki systems are severely power-constrained. Battery operated wireless sensors may need to provide years of unattended operation, often with no way to recharge or replace batteries. Contiki provides a set of mechanisms for reducing the power consumption of the system on which it runs. The default mechanism for attaining low-power operation of the radio is called ContikiMAC. With ContikiMAC, nodes can be running in low-power mode and still be able to receive and relay radio messages.

7.2.6.3   Simulation

The Contiki system includes a network simulator called Cooja (Figure 7.11). Cooja simulates networks of Contiki nodes. The nodes may belong to one of three classes:

- emulated nodes, where the entire hardware of each node is emulated;
- Cooja nodes, where the Contiki code for the node is compiled and executed on the simulation host;
- Java nodes, where the behavior of the node must be reimplemented as a Java class.

A single Cooja simulation may contain a mixture of nodes from any of the three classes. Emulated nodes can also be used, so as to include non-Contiki nodes in a simulated network.

In Contiki 2.6, platforms with TI MSP430 and Atmel AVR microcontrollers can be emulated. Cooja can be very useful because of its emulative functions, which help developers in testing applications. This speeds up the development process: without a simulator the developer would have to upload and test every new version of firmware on real

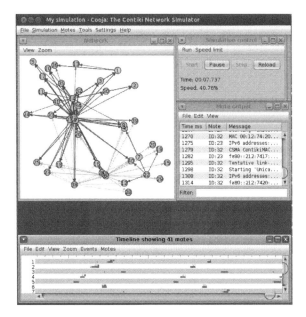

Figure 7.11 Screenshot of Cooja Contiki network simulation for an Ubuntu system with Contiki 2.6 running on 41 nodes forming an IPv6/RPL/6lowpan network.

hardware. This would be a long process, because most motes can only be flashed through a (slow) serial port.

7.2.6.4 Programming Model

To run efficiently on memory-constrained systems, the Contiki programming model is based on protothreads. A protothread is a memory-efficient programming abstraction that shares features of both multi-threading and event-driven programming to attain a low memory overhead. The kernel invokes the protothread of a process in response to an internal or external event. Examples of internal events are timers that fire, or messages being posted from other processes. Examples of external events are sensors that trigger, or incoming packets from a radio neighbor.

Protothreads are cooperatively scheduled. This means that a Contiki process must always explicitly yield control back to the kernel at regular intervals. Contiki processes may use a special protothread construct to avoid waiting for events while yielding control to the kernel between each event invocation.

7.2.6.5 Features

Contiki supports per-process optional preemptive multi-threading, inter-process communication using message-passing events, and an optional GUI subsystem with either direct graphic support for locally connected terminals or networked virtual display with virtual network computing or over Telnet.

A full installation of Contiki includes the following features:

- multitasking kernel
- optional per-application pre-emptive multithreading
- protothreads
- TCP/IP networking, including IPv6
- windowing system and GUI
- networked remote display using virtual network computing
- web browser (claimed to be the world's smallest).

7.3 Vision and Architecture of a Testbed for the Web of Things

With the aim to foster the development and diffusion of the IoT, applications are starting to be built around the well-known web model, leading to the advent of the so-called Web of Things (WoT). The web-based approach has proved to be the driver for the wide diffusion of the Internet and there is a common feeling that this will apply also to the IoT [231]. WoT applications rely on specific web-oriented application-layer protocols, similar to HTTP, such as the Constrained Application Protocol (CoAP) [7] and, more generally, protocols complying with the REpresentational State Transfer (REST) architectural style.

We present here the design and deployment of a heterogeneous and innovative WoT-based application-oriented testbed, called the Web of Things Testbed (WoTT). Its main goal is to allow developers to easily design and evaluate new WoT-oriented services and applications in a real IoT environment and to effectively test human–object interaction mechanisms, which will play a fundamental role in broadening the range of IoT users. WoTT is particularly suited for this purpose because its architecture is completely based on standard protocols and mechanisms; it uses no custom or proprietary solutions that would jeopardize the interoperability among nodes.

The main goals of the WoTT can be summarized as follows:

- to hide low-level implementative details;
- to enhance network self-configuration, by minimizing human intervention;
- to transparently manage, at the same time, multiple protocols and platforms;
- to provide a platform for the design and testing of human-object interaction patterns.

In order to properly test new WoT-related applications, the WoTT consists of several types of node that differ in terms of both computational capabilities and radio access interfaces. Nonetheless, the nodes can be grouped into two main classes: constrained IoT (CIoT) nodes and single board computer (SBC) nodes. CIoT nodes are mainly based on the Contiki OS and correspond to Class 1 devices according to the terminology introduced in RFC7228 [229]. On the other hand, SBC nodes are more powerful nodes, typically running Linux and having multiple network interfaces. These correspond to Class 2 devices. No matter what the actual nodes really are, the standard communication protocols and mechanisms used in the testbed make it able to manage the diversity of nodes seamlessly, thus making it possible to treat each and every node simply as an IP-addressable host. Tables 7.9 and 7.10 show the details of the CIoT and SBC nodes used in the WoTT.

Table 7.9 Constrained IoT nodes in the WoTT.

Constrained IoT nodes				
#	Node	Hardware	OS	Network interfaces
6	TelosB	MCU: TI MSP430F1611 RAM: 10 kB ROM: 48 kB	Contiki	IEEE 802.15.4
20	Zolertia Z1	MCU: TI MSP430F2617 RAM: 8 kB ROM: 92 kB	Contiki	IEEE 802.15.4
10	OpenMote	MCU: ARM Cortex-M3 RAM: 32kB ROM: 512 kB	Contiki	IEEE 802.15.4

Table 7.10 Single board computer nodes in the WoTT.

		SBC nodes		
#	Node	Hardware	OS	Network interfaces
20	Intel Galileo	CPU: SoC X Intel® Quark™ X1000 RAM: 256 MB Memory (SD): 8 GB	[Linux] Debian	IEEE 802.3
5	Raspbery Pi B	CPU: Broadcom BCM2835 ARM11 RAM: 512MB Memory (SD): 8 GB	[Linux] Raspbian	IEEE 802.3/802.11
5	Arduino Yun	Linux environment CPU: Atheros AR9331 RAM: 64 MB ROM: 16 MB	[Linux] OpenWRT	IEEE 802.3/802.11
		Arduino environment MCU: ATmega32u4 RAM: 2.5 kB ROM: 32 kB	Arduino	
4	UDOO	Linux environment CPU: Freescale i.MX 6 ARM Cortex-A9 Dual core RAM: 1GB Memory (SD): 8 GB	[Linux] UDOObuntu	IEEE 802.3/802.11
		Arduino-like environment MCU: Atmel SAM3X8E ARM Cortex-M3 RAM: 100 kB ROM: 512 kB	Arduino	

7.3.1 An All-IP-based Infrastructure for Smart Objects

CIoT nodes are connected, at the physical layer, by IEEE 802.15.4 wireless links, while at the network layer IPv6 is used in combination with 6LoWPAN [230] and RPL (the routing protocol for low-power and lossy networks). CIoT nodes equipped with sensors can act as CoAP servers or clients running Erbium, which is a lightweight implementation of CoAP.

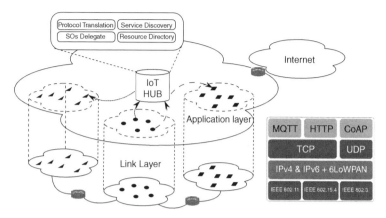

Figure 7.12 The WoTT architecture and protocol stack with emphasis on the role of the IoT Hub in the management of the testbed's heterogeneous network.

As the CIoT nodes, SBC nodes can act as CoAP clients or servers, with fewer constraints than implementations on SBCs. For example, on Arduino Yún nodes, a Javascript application initializes a CoAP server on an instance of node.js. Intel Galileo boards and Raspberry Pi systems can support different languages, ranging from Python to Java.

The WoTT testbed, shown in Figure 7.12, is heterogeneous by design, in order to enable smart objects to communicate seamlessly with each other and to enable communications between the WoTT and external Internet elements (e.g., the cloud, Internet hosts, or consumers). Internet protocol (in particular, IPv6 or IPv6+6LoWPAN) is universally considered a key communication enabler for the future IoT. For this reason, the WoTT adopts IP as a common network substrate, thus allowing for simple integration with the existing Internet world. Figure 7.12 shows the IP-based protocol stack supported by the WoTT. It is worth noting that all nodes in the WoTT use standard protocols at all layers of the protocol stack. As can be seen, several physical (PHY) and MAC protocols (e.g., IEEE 802.11, IEEE 802.15.4, and IEEE 802.3) as well as application layer protocols can be used. The architecture also introduces an innovative network element, called the IoT Hub, which operates at different layers of the protocol stack in order to further enhance the interoperability among communicating devices, by integrating several networks into a single IP-based substrate and, at the same time, by implementing important functions at the application layer. Due to their greater capabilities, in terms of computational power and networking, SBC nodes can be used effectively to implement all the functions of an IoT Hub.

The Wi-Fi networking infrastructure is based on Cisco Connected Mobile Experiences (CMX)[14] and access points, which have already been installed in the WoTT buildings and can be used to track devices (for indoor localization purposes). In particular, the CMX platform provides mobility services REST APIs, which allow developers to enable service customization with location information in mobile applications, such as location-aware equipment tracking, guest access, and device-based services. This feature is currently used to build user-location-aware IoT applications, which can "track and follow" the user throughout the the WoTT building, enabling specific and augmented interactions with the surrounding environment.

Focusing on the application layer, the WoTT currently supports CoAP, MQTT, and HTTP. CoAP is a binary and lightweight web transfer protocol built on top of UDP and follows a request/response paradigm [7]. CoAP has been explicitly designed to work with devices operating in low-power and lossy networks. MQTT is also a lightweight publish/subscribe protocol suitable for constrained smart objects, such as sensors or actuators, and runs on top of TCP. Finally, HTTP is mainly used for communication between the WoTT and external Internet actors or consumers, such as cloud storage services or IoT-unaware clients. We note that, among the WoTT components, the IoT Hub is the key IoT enabler, as it manages the different access technologies and supports full IP connectivity among all objects. By implementing various functionalities at the application layer, all protocols listed in Figure 7.12 can coexist in the same environment.

We also note that the software for smart objects has been developed with different programming languages. This underlines once more that, thanks to the features provided at the application layer, developers can create new IoT applications easily and without any constraint except for compliance with IoT standards.

7.3.2 Enabling Interactions with Smart Objects through the IoT Hub

The WoTT does not simply enable communications between the IoT actors; it represents a "uniform" super-entity, able to provide enhanced

---

14 Cisco Connected Mobile Experiences: http://www.cisco.com/c/en/us/ solutions/enterprise-networks/connected-mobile-experiences/ index.html.

functionalities that go beyond the mere union of the features provided by its components. As shown in Figure 7.12, in order to achieve this holistic vision, the WoTT introduces a new and innovative network element, called an IoT Hub (as presented in Section 6.6), which acts as a gateway that can bridge and merge several networks, using various communication technologies, into a single IP network.

The IoT Hub also implements several functions at the application layer: it manages the services and resources available in the overall infrastructure, thus playing a key role at this layer too. The use of an IoT Hub is expedient for several reasons. The extreme heterogeneity of IoT devices requires mechanisms for their management and the seamless interaction of humans and smart objects. Moreover, focusing on data collection, due to their limited capabilities smart objects may not be able to handle large numbers of concurrent requests, thus making it preferable to limit direct access to them. In other cases, extremely limited devices might act as clients to implement data collection behavior.

Efforts have been made to standardize the design of IoT Hubs. A relevant example is HyperCat [232], which introduces standard specifications to allow servers to expose JSON-based hypermedia catalogues as collections of URLs. Thus, IoT clients are able to discover, using REST methods, data available on servers in HTTPS and JSON format. Unlike the IoT Hub of the WoTT, HyperCat works at a higher level of abstraction, as it is intended to allow reasoning and service composition based on IoT concepts, and does not take into account direct interactions with constrained devices, where specific protocols (e.g., CoAP) should be adopted to minimize energy and memory consumption.

From a networking standpoint, the IoT Hub is a fog node [205], placed at the edge of multiple physical networks with the goal of creating an IP-based IoT network. The IoT Hub plays a fundamental role by implementing the following functions at the link and application layers of the protocol stack:

- *Border router*: the IoT Hub bridges one or more networks (e.g., several IEEE 802.15.4 networks).
- *Service and resource discovery*: the IoT Hub is able to discover the smart objects available in the network and their hosted resources.
- *Resource directory* (RD): the IoT Hub complies with the relevant specifications [14], and maintains a list of all resources available in the bridged networks creating a centralized entry point for applications that need to perform resource lookup.

- *Origin server* (OS): the IoT Hub provides a CoAP server that hosts the smart object's resources.
- *CoAP-to-CoAP* (C2C) *proxy*: the IoT Hub provides proxying capabilities for CoAP requests from external clients targeting constrained nodes.
- *HTTP-to-CoAP* (H2C) *proxy*: the IoT Hub provides HTTP-to-CoAP cross-proxying (i.e., protocol translation) in order to allow HTTP clients to access to CoAP resources.
- *Cache*: the IoT Hub keeps a cache with a representation of most recently accessed resources in order to act as a smart object delegate, minimizing latency and reducing load on constrained devices.

By adopting standard mechanisms and communication protocols, smart objects do not depend on the IoT Hub for their operation; in addition, the IoT Hub is able to mitigate the presence of non-standard components that might not interoperate with standard-compliant devices. It is important to point out that the presence of the IoT Hub is not mandatory for interaction and interoperability, but its presence extends the IoT network and increases its capabilities, by simplifying and hiding complex and important tasks such as service discovery and routing.

7.3.2.1 Integration Challenges

The main challenges encountered in the deployment of the WoTT are related to design: the definition of the different elements and their functionalities, the representation of the different resources and their relationships through suitable hypermedia, and maintenance of compatibility with standards.

The efforts devoted to the design of the WoTT, together with the IoT Hub and the use of standards, has simplified the deployment process, making the integration of all different elements straightforward, notwithstanding their heterogeneity. In particular, the WoTT hides critical implementation issues encountered at lower layers. For example, in a constrained network (e.g., IEEE 802.15.4) the resource advertisement feature is critical, requiring strict assumptions about energy and memory consumption on each constrained device. In Section 4.4, the advertisement of nodes belonging to the network was analyzed, and we derived a multicast solution that allows a reduction in energy consumption. The end users (application-oriented) of the WoTT are not concerned with these implementation details.

Another implementation challenge that is hidden from end users was the creation of IPv4 and IPv6 networks, and the configuration of networks elements (e.g., routers and access points), in order to create a unique IP-addressable network. The firmware provided with common network elements typically does not allow these unusual configurations. To overcome this limitation, the WoTT network components have been flashed with more customizable and advanced firmware, such as Tomato.[15]

7.3.3   Testbed Access and Security

With its goal of becoming a publicly available platform, the WoTT complies with security policies and mechanisms, adopting strong defense measures in order to counteract security threats. These threats may come from external environments, as well as from malicious internal IoT entities. The experience gained from the years of development of the web provides useful security technologies and mechanisms that can be taken as a reference and applied to enhance the protection and reliability of IoT environments.

With access by external clients, the WoTT is able to manage and authenticate request issuers, defining at the same time processing policies and rules. To comply with these specifications, the WoTT allows external connections through VPN and SSH tunnels, using credentials issued by WoTT administrators.

Several security mechanisms for smart object interaction are implemented within the testbed. TLS (in conjunction with HTTP) and DTLS (the secure transport for CoAP) are implemented on a set of devices that host resources that need to be accessed through secured application protocols. The IoT Hub implements both security protocols in order to provide secure resource access through proxying. Of course, end-to-end security through proxying cannot be ensured, as no TLS-to-DTLS mapping is defined.

7.3.3.1   The Role of Authorization

While DTLS and TLS are implemented to enforce confidentiality and authenticity of communication between endpoints, real-world IoT systems must be able to manage multiple users accessing the resources deployed in a smart environment. This is when authorization comes

---

15 Tomato firmware: `http://www.polarcloud.com/tomato`.

into the picture: mechanisms must be defined in order to ensure that only authorized parties can interact with objects. For example, in a multi-occupant building hosting the WoTT, offices should be able to be unlocked only by those who have been granted access and not just to anyone who has discovered a "smart lock". Even though authorization is obviously a critical issue, research has not yet focused on defining mechanisms to grant authorization and manage access policies in IoT scenarios. In order to take this issue into account, the WoTT implements an authorization architecture based on the OAuth protocol, called IoT-OAS (see Section 5.3.2). This architecture implements a lightweight delegation approach to authorization.

7.3.4 Exploiting the Testbed: WoT Applications for Mobile and Wearable Devices

Thanks to the high-level of abstraction and full interoperability of the WoTT, web, mobile, and wearable WoT applications have been developed using it (Figure 7.13). Due to the need to fill the gap between the mobile/wearable world and the IoT, it is important to define innovative interaction patterns, through which developers might build, deploy, and connect their applications so that these can be used in a simple way by end users. To validate the benefits of the WoTT and demonstrate the ease of integration of a newly deployed application with the testbed, wearable-enabled and mobile-oriented applications have been

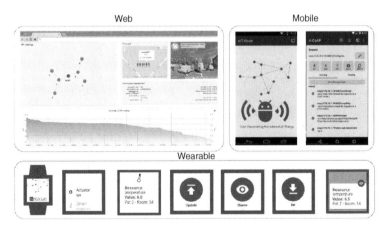

Figure 7.13 Web, mobile, and wearable REST-based testbed applications.

developed with the goal to start interacting with "things" in a natural way, similar to what a person experiences everyday on the web.

Thanks to their portability, wearable and mobile devices are attractive ways to track people's actions in the real world. Indoor localization features have been developed in the WoTT, using triangulation techniques implemented through the CMX access points. The availability of localization APIs in the Cisco CMX system allows applications to be built that take into account the location and movement of users throughout a building. This is an extremely interesting feature, which can pave the way for development of applications that can actually follow the user and even anticipate their movements in the building. Together with access-point-based localization, the use of on-board inertial measurement units can further improve the tracking capability.

Mobile devices play an important role in our architecture. Besides interacting with the deployed smart objects, they can also act as smart objects, providing data generated by their onboard sensors. Mobile devices can thus be considered as WoTT nodes, making the WoTT a highly dynamic and evolving system.

By providing a uniform and application-oriented platform, the WoTT can be used effectively by developers to create and test real-world IoT applications easily and in short time, thus making the WoTT more attractive than other currently available platforms. This is made possible by the ready-to-use capabilities and the direct/active interactions that a deployed application may have with the resources availables into the WoTT. From an operational point of view, developers only have to run their applications involving testbed resources, without any additional need to virtualize environment services, and becoming part of a WoT scenario, in which consumers are not only "readers", but active participants.

Based on the capabilities provided by the WoTT, we have developed a mobile and wearable application (Figure 7.13), which recreates the basic steps envisioned for a WoT application that can really be used by anyone. These applications have been tested on the Android Wear platform using LG G watches and Android 5.0.1 smartphones. In the near future, as more and more smart objects are deployed, vendor-provided apps will no longer be the typical means to interact with things: a more standard approach will be needed in order to do so effectively.

The developed application performs the following steps. The mobile device first discovers smart objects in its proximity, proactively and reactively, by means of standard service discovery and resource directory mechanisms, and then forwards the collected information to its connected wearable device interface. Through wearable interfaces, a user can see and browse a list of all the resources that have been discovered and select one to interact with. Interactions are performed according to the function set specified by the selected smart object (e.g., a light bulb might provide an on/off switch, while a temperature sensor might just provide a way to read its value). Resources and interactions are revealed gradually, according to the REST paradigm, so that the application can adapt itself dynamically.

We note that WoTT resources can be deployed on different platforms (e.g., different smart objects) and using heterogeneous protocols. However, this is completely transparent to a developer, who can access all these resources, only constrained by the need to use standard protocols. This possibility is granted by the IoT Hub's abstraction ability and is not a feature of other existing testbeds (e.g., SmartSantander).

A user can interact with smart objects through one of the following approaches:

- *Polling* allows the user to retrieve the value associated with the queried resource, by performing a CoAP GET request.
- *Observing* can be used to receive asynchronous updates when the value of the specified resource changes, adopting an efficient mechanism and avoiding the need for periodic data polling.
- *Acting* is used to set the value of a specified resource, depending on the function set provided by the resource.

The observing approach, which was not defined in HTTP, is a lightweight CoAP-oriented interaction mechanism [9]. Resource observing is achieved by ussuing a CoAP GET request containing an Observe option. This option instructs the target smart object to add a new subscriber, which will receive, in push mode, subsequent resource updates. Subscribers also have the ability to stop observing a resource at any time, by unsubscribing from updates.

7.3.5 Open Challenges and Future Vision

The use of standard protocols, the use of well-known web-based approaches, and the widespread diffusion of hardware platforms are changing and opening the IoT world to new developers, businesses, and users. Against this dynamic and evolving background, the availability of real and accessible resource-oriented testbeds that allow active and direct interaction with low-level nodes and services is one of the key enablers for the future adoption of the IoT by a broad audience, and even for the final transition from the IoT to the WoT.

The ongoing WoTT experience provides a concrete example of new architectural and networking solutions that address important open issues and challenges for the IoT. The experience gained in designing and implementing our testbed has confirmed that there is a concrete need for open evaluation platforms aimed at integrating and experimenting with innovative solutions, in order to develop applications that can really bridge today's gap between users and things. The heterogeneity (in terms of hardware and software) of our testbed, testifies to how the proper use of standard protocols (e.g., HTTP, CoAP, and MQTT) and interaction paradigms and approaches (such as REST and service/resource discovery), are fundamental to creating transparent and dynamic interactions among multiple smart objects and personal user devices, such as smartphones and wearables. Moreover, this heterogeneity can be extended, with limited effort, by improving the IoT Hub features (e.g., adding support for Bluetooth devices). This would open up the WoTT to a new emerging category of IoT-enabled devices, namely Bluetooth Low Energy, thus allowing its integration with other service discovery mechanisms, such as UriBeacon.[16]

Nevertheless, important open issues still need to be tackled to bring the IoT into our daily life. Security, interoperability, data processing, human interaction, and easy development and deployment are the keywords that will characterize academic and industrial R&D activities over the next few years. IoT-application-oriented testbeds represent the perfect experimental playgrounds to boost and support development of IoT applications, creating a common space that developers, hardware manufacturers, and companies can exploit in order to reach the goal of making the IoT accessible and easy to use for everyone, just as the Internet is right now.

---

16 The UriBeacon Open Standard, (Apr 2015) `http://uribeacon.org`.

7.4 Wearable Computing for the IoT: Interaction Patterns with Smart Objects in RESTful Environments

Over the next few years, the IoT is expected to become a reality, merging the social world, the physical world, and the cyber world to enable new applications and forms of interactions between humans and connected smart sensing/actuating devices. As billions of smart objects will be deployed in the environment, users should be able to discover and interact with objects in their proximity in a seamless and transparent way. While smartphones have become an extremely popular computing device, the advent of smart wearable devices, such as Google Glass and the Apple Watch, has provided even more effective means to bridge the gap between humans and smart objects. In this section, we analyze the characteristics of wearable applications for IoT scenarios and describe the interaction patterns that should occur between wearable/mobile devices and smart objects.

7.4.1 Shaping the Internet of Things in a Mobile-Centric World

In recent decades, the Internet has allowed people to access and consume services on a global scale, using traditional hosts and always-connected mobile devices, such as their smartphones, typically using the World Wide Web. By connecting objects and devices, the IoT will fully exploit the potential of networking and enable innovative services to be offered in a wide range of scenarios, such as home and building automation, smart cities, and healthcare, integrating new paradigms for human-to-machine (H2M) and machine-to-machine (M2M) interactions. Many IoT applications will allow people to interact with smart environments, allowing them to obtain information and to change the environment according to their needs and preferences, sometimes without actually being in the loop. IoT applications will take advantage of the wide diffusion and pervasive deployment of smart objects: tiny devices equipped with a microcontroller, a communication interface (wired or wireless), a power supply, and a set of sensors and/or actuators to be used to interface with the surrounding environment.

Several players are developing innovative IoT-related products in a variety of fields, which are starting to reach end-users, who are now becoming aware of the integration of physical and cyber worlds. The "gold rush" of the IoT-era, driven, on one hand, by developers' will to demonstrate the feasibility of interconnecting everyday devices to people and, on the other hand, by their hope to turn their custom solutions into standards for the general public, has created a plethora of closed vertical solutions. This is leading to a highly fragmented market: a babel of incompatible solutions, as opposite to a highly standardized and interoperable environment, which is what the Internet (of Things) should be [231]. In order to prevent the IoT reaching a dead end due to fragmentation caused by these vertical solutions, much effort has been expended in research projects and standardization organizations, such as IEEE, IETF, and the IPSO Alliance.

The successful model for communication represented by the Web has been considered as a reference point for the IoT as well. We can now assume that the IoT will be a network of heterogeneous interconnected devices; this will be the infrastructure for the WoT.

Mobile devices (smartphones and tablets) have become the most popular computing devices in the world, topping the per-capita rate of penetration of personal computers for the first time in mid-2012 [233]. We are living in a mobile-centric world, characterized by symbiosis and interdependence between users and their mobile devices, to the point where one cannot work without the other. Thanks to their rich capabilities, mobile devices are handy and ready-to-use gateways to all IoT objects deployed nearby or far away: a cyber-"Swiss-army knife" for the IoT. This ease of access be enhanced further by the evolution of wearable computing, which has the potential to augment the ways people interact with services (e.g., through voice control, gestures, or touch) or provide information to services (e.g., heart rate monitoring, fitness applications). The smartphone is the enabler for a user who wants to control and interact with smart objects in his proximity and wearable devices. Interaction with smart objects through wearable devices will represent a quantum leap towards the full integration of the social world, the cyber world, and the physical world, and will be, in fact, a milestone for the widespread adoption of the IoT. The interaction between wearable devices and IoT objects should occur through well-defined patterns, which must take into account their nature, while providing maximum usability and the best user experience.

7.4.2   Interaction Patterns with Smart Objects through Wearable Devices

The evolution of mobile and wearable computing has changed the way people use online services. They are now always connected, whether at home or on the go. In this context, there is a concrete need to fill the gap between mobile devices and the IoT. A paradigm shift in specific aspects is needed to let people access and use the IoT with the same simplicity as when accessing the Internet and, possibly, to enable new and more natural forms of interactions, which will broaden the range of IoT users.

7.4.2.1   Smart Object Communication Principles

The idea of an IP-based IoT has been around for years and is now considered as a fact. The adoption of the IP protocol for smart object addressing and communication is the driver to full interoperability and integrability of the IoT with the existing Internet. In particular, the use of IPv6 has been foreseen as the solution to the management of billions of globally-addressable devices. A fundamental principle that has driven IoT research and the work of standardization institutions is the maximization of the reuse of standard Internet protocols and mechanisms. Due to the limitations of low-performance IoT devices, it is not always possible to use the traditional TCP/IP protocol stack. New energy-efficient and low-overhead communication protocols and data formats, which take into account group communication, mobility, and interactions among multiple devices, have been designed. The introduction of 6LoWPAN has solved the problem of bringing IP to low-power devices. Thanks to the definition of common standards, the IoT is ready to reach the next level: the Web of Things (WoT).

The Web is by far the most popular and familiar interaction model on the Internet. The WoT is being designed around well-known concepts and practices derived from the Web, such as the REST paradigm. The REST paradigm was introduced to loosely couple client applications with the systems that they interact with and to support long-term evolution of systems and provide robustness. At the application layer, CoAP has been designated as the standard protocol for the WoT, similar to the role of HTTP for the Web. In fact, CoAP has been designed to work with HTTP, to which it maps easily for integration purposes, by inheriting from it the identification of resources through URIs, the request/response communication model, and the semantics of its methods. However, it is not possible to fully map CoAP to HTTP for a number of reasons. In particular, CoAP is a binary UDP-based protocol, while HTTP implies connection-oriented communications. CoAP also introduces some significant enhancements and optimizations that are relevant for low-power devices, such as support for group communication and resource observation. This latter feature allows client applications, after an initial request, to receive updates for a given resource in successive response messages, without the need to perform periodic polling. Resource observation introduces a new communication paradigm within REST. Note that REST is not the only communication pattern that is ever going to be adopted in the IoT, similarly to what happens on the traditional Internet: protocols using publish/subscribe mechanisms fit well when a single source delivers information to multiple receivers.

7.4.2.2   Interaction Patterns

A smartphone is required when using wearable devices: it is responsible for performing all the heavy duty and complex tasks (notably processing and communication). Wearable devices are, in fact, extensions of a mobile devices, the latter acting as central hubs. In order to enable truly seamless and practical interaction with smart objects through wearable devices, it is crucial that the smartphone takes care of these tasks behind the scenes. In smart environments, wearable devices will thus play the role of true enabler of natural interactions with things, which is what people expect. As pointed out by Sergey Brin in a TED talk at the time of the launch of the Google Glass, the use of smartphones isolates people, making them constantly and obsessively looking down at their devices [234]. The Google Glass (and, in general, any wearable device) should let people keep their heads up, instead, thus allowing for simpler and more natural interactions. This change is going to have an even bigger impact when users start interacting with a myriad of smart objects, rather than just texting or checking emails.

Wearable devices can be classified into two categories. Passive wearables do not require any direct human interaction (e.g., heart rate monitors and step counters) and are therefore tightly coupled with a mobile device hosting a custom app to control them. Active wearables (e.g., smart watches and glasses) provide information to users and can be used as sources of input, for control of the wearable itself and also the mobile device they are connected to. As a consequence, they

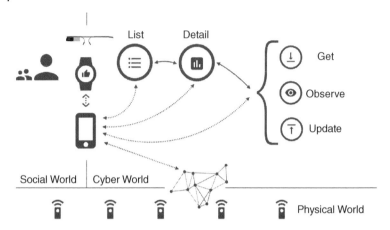

Figure 7.14 Interaction patterns between wearable devices and smart objects.

can extend their reach to other devices, which can be located in their proximity or at a distance. For our purposes, we focus exclusively on active wearables, as they should be the drivers of interactions with smart objects.

Figure 7.14 shows the interaction patterns that we envision between humans, using wearable devices, and smart objects that may be around them. As described above, smartphones and wearable devices are the bridge between the social world (users), the cyber world (networked hosts), and the physical world (connected smart objects). Typically, there are some questions that users might ask when trying to interact with a smart environment:

1) What objects are around me?
2) Do I have the right privileges to control or interact with it directly?
3) What is a given object? What can it do and what can I do with it?
4) How can I interact with it?

In order to answer the above questions and provide a natural sequence of interactions, we have envisioned an operational flow, which can be summarized as follows:

- The smartphone discovers the smart objects in its proximity (Q1) for which the user has been granted the appropriate access privileges (Q2), by means of suitable service discovery (e.g., ZeroConf, as presented in Section 4.3, or low-power radio beacon broadcasting) and resource discovery (e.g., CoAP resource directory) mechanisms, and presents this information on the wearable device interface.

- The user can select one item (using vocal input or touch) and get detailed information related to the selected resource, combining the information retrieved in the previous step with other descriptors that can be retrieved on the fly, such as the function set that can be used to operate the resource [55] or a custom user interface (Q3).
- The user can then browse through the available forms of interaction with the selected resource, through specialized interfaces for the possible actions that can be performed (Q4).
- The user interacts with the resources, always using the smartphone as a bridge to perform the actual communication with smart objects, adopting one of the following methods:
  - *Get*: this method allows the user to retrieve the value associated with the resource, by performing a CoAP GET request (e.g., to check the status – closed or open – of a smart-lock-equipped door).
  - *Observe*: this method employs an efficient mechanism for receiving asynchronous updates when the value of a resource changes, thus avoiding having to periodically poll for data [55]. This is achieved by performing a CoAP GET request containing an Observe option to instruct the object to send updates. An observe relation can be then torn down at any time (e.g., to monitor the temperature of a room so that a thermostat can be activated).
  - *Update*: this method is used to act upon a resource to set its the value, by performing a CoAP POST or PUT request, depending on the function set provided by the resource (e.g., to switch lights on or of).

The use of CoAP is not strictly required: the presence of cross-proxies, which might perform protocol translation, can enable the same interactions with CoAP-unaware clients that may use HTTP.

7.4.3   Implementation in a Real-world IoT Testbed

Following the interaction patterns described above, a wearable-based application for interacting with smart objects has been implemented and tested within an IoT testbed deployed inside our department, with 70 IP-addressable devices running CoAP servers that host sensing and actuating resources. The IoT testbed consists of several types of nodes, which differ in terms of computational capabilities and radio access interfaces. The considered nodes can be grouped into two main classes: 36 constrained IoT nodes and 34 single board computer (SBC) nodes.

The constrained IoT nodes are Class 1 devices, according to the terminology introduced in RFC 7228 [229], based on the Contiki operating system and using IEEE 802.15.4 radio. Specifically, the following constrained IoT nodes were used:

- 6 TelosB[17]
- 20 Zolertia Z1[18]
- 10 OpenMote-CC2538.[19]

The SBC nodes are Class 2 devices, typically running a Linux operating system and with multiple network interfaces. Specifically, the following SBC nodes were used:

- 20 Intel Galileo[20]
- 5 Raspberry Pi Model B[21]
- 5 Arduino Yún[22] (dual Linux/Arduino environment)
- 4 UDOO[23] (dual Linux/Arduino environment).

The IoT testbed is purposely heterogeneous in order to simulate the expected real-world IoT conditions, characterized by an extremely high degree of diversity of smart objects. It also highlights how the use of standard communication protocols and mechanisms, combined with our envisioned operational flow, can enforce interoperability and effective interactions among applications, in a real WoT fashion.

The mobile/wearable application was developed on the Android Wear platform using LG G watches and Android Lollipop smartphones. Figure 7.15 shows screenshots of the application

---

17 http://www.memsic.com/userfiles/files/Datasheets/WSN/ telosb:datasheet.pdf.
18 http://www.zolertia.com/ti.
19 http://www.openmote.com/hardware.html.
20 http://www.intel.com/content/www/us/en/do-it-yourself/ galileo-maker-quark-board.html.
21 http://www.raspberrypi.org/products/model-b/.
22 http://arduino.cc/en/Main/ArduinoBoardYun?from=Products .ArduinoYUN.
23 http://shop.udoo.org/eu/product/udoo-dual-basic.html.

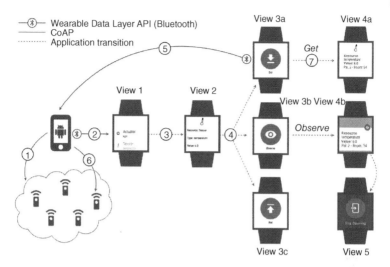

Figure 7.15 Mobile/wearable application for interaction with smart objects: operational flow with indication of communications and screenshots of the implemented Android application.

running on the smartwatch, and how these map to the previously described flow.

There are essentially three actors involved in the application's lifecycle: the smartphone, the wearable device, and the smart objects. Each operation, which may be initiated by any of these, involves an information exchange that affects them all. The smartphone is the gateway between the smart objects and wearable devices, as shown in Figure 7.14. The communication between the smartphone and the smart objects occurs using CoAP, which is based on the Californium library [94]. The smartphone application (SA) communicates with the wearable companion application (WA) using the Android Wearable Data Layer API, which uses Bluetooth for communication. At startup (step 1), SA performs resource discovery using a combined mechanism:

- using ZeroConf, it discovers CoAP servers, whose lists of resources are then retrieved by performing a GET request to the /.well-known/core URI;
- a multicast GET request to the /.well-known/core URI is sent to all smart objects in the network, which return their lists of hosted resources.

After the SA has gathered the list of all available resources, it pushes this information to the WA (step 2), which presents the list of resources to the user (View 1). The user selects a resource of interest, and is presented with a dedicated detail view of that resource (step 3), which includes resource-specific information, such as its human-readable name and type (View 2). According to the interface of the resource [55], the WA presents (in a 2D picker) swipeable cards that contain action buttons, which can trigger interactions with the resource, using the previously described Get, Observe, and Update schemes (Views 3a, 3b, and 3c, respectively). Upon user selection (for instance, in Figure 7.15, a Get interaction is selected), a message is sent by the WA to the SA (step 5), which then issues the corresponding CoAP request and receives the response from the smart object (step 6). The response is then processed and the results are pushed to the WA (step 7) and then presented to the user (View 4a). In case of resource observation (View 3b), when the smart object sends updates to the SA, the SA pushes this information to the WA, which presents a notification (View 4b). A dedicated view is also available to the user to allow them to stop observing resources (View 5).

7.4.3.1 Future Vision: towards the Tactile Internet

The strong evolutionary trend in several ICT fields, from electronics to telecommunications and computing, has brought significant innovations in the way people use and interface with the environment. On one hand, mobile and wearable computing have allowed people to be connected anywhere and at anytime and to consume services while on the go. On the other hand, cheap embedded systems and low-power operating systems and communication protocols have paved the way to the development of the IoT. Although these fields have evolved independently, they are converging and complement one another more and more, thanks to the flexibility and inherent capabilities of smartphones. The combination of and integration between the IoT and mobile/wearable devices is allowing industry and researchers to envision a path towards a true Internet of Anything [235], capable of connecting everyone and everything.

Improvements in communication technologies, in terms of ever-increasing speed, reliability, and security, are paving the road to the delivery of groundbreaking applications to the general public. While the IoT is being designed around low-power devices with limited capabilities, the enhancements that future 5G networks are going to bring about will foster the development of services and applications that will no longer suffer from the strong limitations of earlier technologies, such as 4K video streaming or real-time remote control. Combined with fog computing techniques [205], aiming at minimizing latency and enabling real-time applications by moving some processing load to the edge of local networks, the idea of end-to-end communication in times of the order of milliseconds is going to become a reality. These technologies will lead to the ability to perform haptic (meaning touch-based) interactions with connected things, either in the proximity or at a distance, thus bringing about the rise of the "Tactile Internet" [236] and its application to healthcare, education, robotics, and many others fields. Wearable devices will play a significant role in this mission and are already set to be the drivers of many kinds of haptic interactions with things.

Thanks to the IoT, users will no longer need to be in charge of monitoring and controlling things when this is not needed. However, they will still be able to interact with their surroundings when they want to do so. Natural interaction patterns between users and smart objects through wearable devices will be a key enabler to reach this goal: these models will guide and teach users by increasing their awareness without requiring them to learn new paradigms.

7.5 Effective Authorization for the Web of Things

By connecting billions of "smart objects", the IoT is set to have a dramatic impact on how people interact with the surrounding environment, whether at home or in public places. With forecasts of such a gigantic number of deployed smart objects, much attention is being given to the societal and economic impacts of the IoT.

Technologies such as Apple's HomeKit[24] or the new Google Brillo/Weave duo,[25] are just two examples of how big players have approached the IoT; that is, by trying to affirm proprietary solutions as standards, just as in the early days of the Internet. In order to tame the extreme diversity of smart objects, in terms of processing capabilities, communication technologies, operating systems, and

---

24 https://developer.apple.com/homekit/.
25 https://developers.google.com/brillo/.

implemented functions, in recent years much effort has been dedicated to defining standard mechanisms and protocols that will actually enable things to be connected to each other and with the Internet. Moreover, the experience gained in the design of the traditional Internet is a valuable asset that can be exploited in order to avoid making the same errors and seeing the same issues as encountered over the last 25 years of the Internet.

From a business standpoint, the time is ripe for IoT products to reach the market. However, there is still much effort required to make the IoT simple to use, practical, and effective, in order to broaden the range of users and to show the real capabilities of what the IoT can do, instead of putting on the market products that are attractive only to specialists and experts. The expectations and concerns of users about new products are primarily related to the possibility to interact with smart devices in a seamless and effective way and to do so securely. While tasks that can be achieved with a certain product or application are business matters that go beyond the domain of pure technology, there are requirements that must be met to ensure that the IoT will not head into a dead end.

The Web of Things (WoT) paradigm pushes interoperability further, by letting applications interact using standard and uniform interfaces, with the same simplicity experienced on the web. Besides RESTfulness, the WoT will also have additional requirements, such as support for different communication paradigms including one-to-many communication, and support for publish-subscribe patterns [237]. Thanks to these standardization efforts, we can consider interoperability more as a business issue rather than a technological one.

In the future, users (and things) will be dealing with a the large number of smart devices deployed either in private or public spaces. Even if interoperability is ensured, it is not feasible to manually configure each and every device in order to discover which objects are nearby and which functions they provide. In order to let applications be ready to operate with devices with minimal (or no) configuration, mechanisms for the discovery of devices, services, and resources have been proposed, as presented in Chapter 4. These functions may also be implemented on specific network elements, such as IoT Hubs (see Section 6.6), which can act as communication enablers and provide discovery as a service.

Applying Web technologies to the physical world, leading to the "Physical Web"[231], will enable simple interactions with and among things. In this context, the concerns of users about new products are primarily related to secure access; that is, only parties with proper authorization must be able to interact with smart objects. However, in the rush to launch IoT products, security and authorization issues have not been taken into proper consideration from a standardization and long-term perspective. Secure and authorized access to smart objects has typically been provided through product-related cloud platforms, which mediate between a smartphone app and the smart object. Notwithstanding this, standard and effective mechanisms for managing authorized access to things are required to really make using and sharing things with others simple and safe. In this section, we present a standards-based authorization framework for WoT applications, which allows users to effectively define fine-grained policies to access shared resources. The proposed framework relies on the IoT-OAS architecture (Section 5.3.2), which implements a delegation strategy for authorization. The use of IoT-OAS makes it feasible to let smart objects implement effective access policies to hosted resources and to minimize the amount of processing load. There are also other benefits, such as

- remote management of access policies without direct intervention on smart objects;
- extension of smart object battery lifetimes;
- scalability, in terms of the number of external clients that can access resources.

IoT-OAS is suitably extended by integrating a simple set of messages that can be used to manage ownership and shared access to resources, with minimal operation required by the actors involved. An implementation of the complete framework is presented here in order to show the simplicity of the proposed approach and to highlight all the benefits that it can introduce.

In the WoT, besides user authentication, things will need to authenticate in order to communicate securely. Password-based authentication, which works fine with humans, does not apply well to things: embedding confidential information in (possibly unattended and vulnerable) objects might not be the safest option. Moreover, for constrained devices, traditional encryption algorithms and security protocols used on the Internet, such as TLS, might not be feasible. Lightweight cryptographic algorithms, such as PRESENT, TEA, and SEA, with shorter keys, have been proposed to provide security for these application contexts. At the same time, lightweight secure transports, such as DTLS (used in conjunction with CoAP, an arrangement denoted as CoAPs), might be preferred over their connection-oriented equivalents.

Authentication alone is not enough to provide sufficient safety and security in the IoT, where the cyber world and physical world will merge. Authorization issues must be taken into account to ensure that access to sensing and actuating devices, both in private and public spaces, occurs only by authorized parties. The Delegated CoAP Authentication and Authorization Framework (DCAF) [193], proposed by the IETF Authentication and Authorization for Constrained Environments (ACE) Working Group, defines a protocol to delegate the establishment of a DTLS channel between resource-constrained nodes and external clients. DCAF targets scenarios where CoAP requires a secure transport. Even though the delegation approach is expedient to reduce the processing load and requirements on constrained nodes, DCAF does not take into account authorization aspects that are needed when third-parties access resources.

A reference approach to third-party access to resources is OAuth [173, 175]. OAuth has been used effectively for online social networks that provide access through RESTful APIs. However, OAuth is based on HTTP, which cannot always be used as a communication protocol with smart objects.

When referring to authorization in the WoT, there are complex problems that need to be solved. Object owners might want to restrict access to their resources, while still being able to grant access privileges to authorized parties (e.g., a person might want to share access to a home smartlock with their spouse, but not with any other individual). These problems can be categorized as:

- owner-to-owner authorization (when the object owner authorizes himself to control an object);
- owner-to-any authorization (when the object owner authorizes other parties to control an object he owns) [237].

Owner-to-owner authorization is addressed by OAuth 2.0 [175]. However, while and the user-managed access profile [238] of OAuth 2.0 addresses owner-to-any authorization in the web, it does not suit constrained environments. Owner-to-any authorization in the WoT will be an important and common scenario and needs to be addressed effectively.

7.5.1 Authorization Framework Architecture

The proposed framework builds on and completes the IoT-OAS architecture, introducing mechanisms for the issuing of access tokens that can be used to access resources in IoT applications. The IoT-OAS architecture introduces a delegation approach to authorization that can be used effectively to manage and enforce fine-grained access polices for any kind of resource, either in the Web for in the WoT. IoT-OAS provides a *verify()* function that can be invoked by smart objects or intermediate network elements, such as proxies, called service providers, that need to check incoming requests from service consumers before serving them, as shown in Figure 7.16. The presence of IoT-OAS is transparent to requesting parties, which can therefore operate without any change required in their implementation.

7.5.1.1 System Operations

The system comprises four different actors:

- SO: the smart object, characterized by an identifier (UUID), which hosts some resources;
- u_1: the owner of SO;
- IoT-OAS: the delegation authorization service;
- u_2: a user, who does not own SO, but is going to control it.

Users can log on to IoT-OAS using any suitable authentication mechanism it accepts, such as OpenID, Facebook login, or Google. Upon login, users are assigned an access token to be used for any interaction with IoT-OAS. We assume that u_1 and u_2 already have accounts on IoT-OAS. Alternatively, accounts can be created on the fly by linking to Google or Facebook accounts.

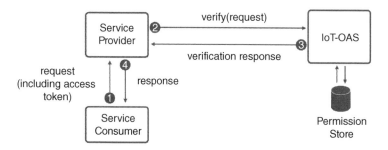

Figure 7.16 Verification of access grant with IoT-OAS.

Table 7.11 Authorization framework message specification.

API	Input parameters	Description
register	Device UUID	Add a new device to the list of owned devices
request	Device UUID Set of permissions	Request to gain permission to control someone else's object
grant	User Device UUID Set of permissions	Share an object with another user

The framework extends IoT-OAS by providing a set of messages that users can exploit to interact with IoT-OAS in order to manage and request access grants to smart objects. All API calls are authorized by IoT-OAS with OAuth, using the access tokens received upon login. The main methods in the framework on IoT-OAS are detailed in Table 7.11.

In order to give the finest granularity in expressing the access privileges granted, a permission is defined by the tuple $\langle res, act, exp \rangle$, where *res* is the URI of the resource, *act* is the REST method to act on the resource, and *exp* is the expiration time of the grant. The framework takes into account the following three operational cases:

- owner-to-owner authorization;
- reactive owner-to-any authorization: u_2 asks permission to u_1 to access an object owned by u_1;
- proactive owner-to-any authorization: u_1 grants permission to u_2 to access an object he owns.

The use of the API to interact with IoT-OAS allows for definition of a simple protocol to implement the three operational cases easily.

Owner-to-owner Authorization

The owner of an object (e.g., a someone buying an appliance for the home or a system administrator for a company), denoted as u_1, needs to authorize himself to gain control of the object. We assume that u_1 has already logged on to IoT-OAS. The login procedure results in the issuing of an access token that u_1 will include in all requests to be identified. The message flow for this case is shown in Figure 7.17a.

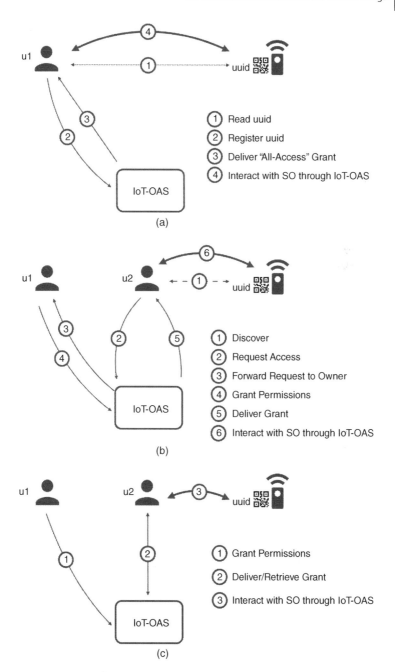

Figure 7.17 Specification of message flows for authorization use cases: (a) access to resources for smart object owner; (b) reactive sharing of smart objects; (c) proactive sharing of smart objects.

Use u_1 can read the UUID of the smart object in some way, such as by scanning a QR code. u_1 sends a *register*(IUUID) request to IoT-OAS. Upon verifying the message, IoT-OAS binds the device (and, subsequently, any resource it hosts) to its owner and, at the same time, invalidates any further registration request for the same UUID. In addition, IoT-OAS adds "all-access" permissions to u_1's access token. u_1 is now authorized to perform any operation on the smart object. All requests targeting the smart object must include the access token, which will be used to validate the request to IoT-OAS's *verify()* method.

Reactive Owner-to-any Authorization

Suppose a user u_2 discovers that a smart object owned by u_1 is in his proximity and wishes to use it. u_2 must be granted a set of permissions to access the smart object's resources. We assume that u_1 and u_2 have already logged on to IoT-OAS. The message flow for this case is shown in Figure 7.17b. Let P be the set of permissions that u_2 requires to access the smart object. To receive the grant, u_2 sends a *request*(UUID, P) message to IoT-OAS, which forwards the message to the owner of the smart object, u_1. u_1 receives the message and grants a set of permissions P^g, which may differ from P. u_1 sends a *grant*(u_2, UUID, P^g) message to IoT-OAS, which adds the set of permission to u_2's access token and notifies u_2 of the result of the operation that he initiated. By including his token in all requests, verified by IoT-OAS, u_2 will be able to access the smart object's resources.

Proactive Owner-to-any Authorization

Suppose u_1 wants to share the smart object with u_2, without u_2 explicitly soliciting permission. We assume that u_1 and u_2 have already logged on to IoT-OAS. The message flow for this case is shown in Figure 7.17c and is a simple subset of the operations that occur with the reactive case. u_1 sends a *grant*(u_2, UUID, P) message to IoT-OAS, which adds the set of permissions to u_2's access token. At this point, depending on the specific application implementation and/or user settings, there are two ways u_2 can receive the grant:

- using a suitable push notification service, IoT-OAS notifies u_2;
- as in the reactive case, after sending a *request*(UUID) message, u_2 will be notified immediately.

7.5.2 Implementation and Validation

Following the definition of the authorization framework, a complete implementation of the proposed architecture has been realized to validate the approach with real smart objects and personal devices. Communication occurs using CoAP/CoAPs, based on the Californium library [94] and its DTLS 1.2 module Scandium.[26] An Android application as implemented to support the user operations defined by the authorization flow. At startup, the user can authenticate through Facebook or Google+, as shown in Figure 7.18a. After the user has logged in, they can view the list of owned and shared devices and the associated permissions, as shown in Figure 7.18b. By clicking on the floating "add" button, the user can register a device they own and that they want to manage, by scanning the QR code associated with it. Figure 7.18c shows the device's management view, which allows the user to interact with available resources and manage usage permissions with other identified consumers.

The application has been integrated with and tested in our company's IoT laboratory, enabling us to consider different scenarios with multiple active users. Through the application, office employees can, at the same time, interact with available sensors owned by the

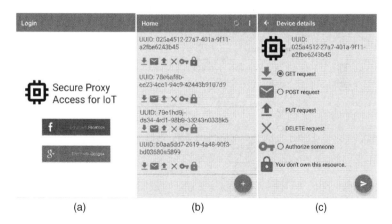

(a)   (b)   (c)

Figure 7.18 Main sections of the Android application: (a) platform authentication with Facebook and Google+; (b) list of owned and shared devices; (c) detail view of a specific device with related permissions.

---

26 https://github.com/eclipse/californium.scandium.

company system administrator (e.g., temperature, light, and humidity) and personal devices on their desk, represented by Philips Hue[27] light bulbs. The meeting room was identified as a common and shared area where, according to appointment schedules, users can dynamically receive access grants from the Sys-admin in order to interact with the room's equipment – items such as smart plugs and wireless lighting – during the meetings.

As the IoT and its applications are expected to progressively become an integral part of our lives, security issues must be addressed in order to cope with the concerns that might prevent the spread of IoT products. In general, people will interact with smart objects that they own and with others that they do not own but that can be shared with them. Authorization mechanisms must therefore be introduced in order to take into account the possibility of users wanting to interact with objects that they do not own. In this section, we have presented a standards-based authorization framework for WoT applications, to effectively enforce fine-grained policies to access and share IoT resources. The proposed framework allows for a simple and seamless management of access permissions to smart objects. Operation flows, relying on the IoT-OAS architecture, have been defined for owner-to-owner and owner-to-any authorization grants. Moreover, a complete implementation of the authorization framework has been realized in order to validate compliance with functional requirements and ease of use.

---

27 http://www.developers.meethue.com/.

References

1 Leiner, B.M., Cerf, V.G., Clark, D.D., Kahn, R.E., Kleinrock, L.,
Lynch, D.C., Postel, J., Roberts, L.G., and Wolff, S. (2009) A
brief history of the internet. *SIGCOMM Comput. Commun.
Rev.*, **39** (5), 22–31. URL `http://doi.acm.org/10.1145/
1629607.1629613`.

2 Fielding, R., Gettys, J., Mogul, J., Frystyk, H., Masinter, L.,
Leach, P., and Berners-Lee, T. (1999) Hypertext Transfer Protocol – HTTP/1.1, RFC 2616. URL `http://www.ietf.org/
rfc/rfc2616.txt`, obsoleted by RFCs 7230, 7231, 7232, 7233,
7234, 7235, updated by RFCs 2817, 5785, 6266, 6585.

3 Fielding, R. and Reschke, J. (2014), Hypertext Transfer Protocol (HTTP/1.1): Message Syntax and Routing, RFC 7230. URL
`http://www.ietf.org/rfc/rfc7230.txt`.

4 Belshe, M., Peon, R., and Thomson, M. (2015), Hypertext Transfer
Protocol Version 2 (HTTP/2), RFC 7540. URL `http://www
.ietf.org/rfc/rfc7540.txt`.

5 Rosenberg, J., Schulzrinne, H., Camarillo, G., Johnston, A.,
Peterson, J., Sparks, R., Handley, M., and Schooler, E. (2002),
SIP: Session Initiation Protocol, RFC 3261. URL `http://www
.ietf.org/rfc/rfc3261.txt`, updated by RFCs 3265, 3853,
4320, 4916, 5393, 5621, 5626, 5630, 5922, 5954, 6026, 6141, 6665,
6878, 7462, 7463.

6 Stewart, R. (2007), Stream Control Transmission Protocol, RFC
4960. URL `http://www.ietf.org/rfc/rfc4960.txt`,
updated by RFCs 6096, 6335, 7053.

7 Shelby, Z., Hartke, K., and Bormann, C. (2014), The Constrained
Application Protocol (CoAP), RFC 7252. URL `http://www
.ietf.org/rfc/rfc7252.txt`, updated by RFC 7959.

Internet of Things: Architectures, Protocols and Standards, First Edition.
Simone Cirani, Gianluigi Ferrari, Marco Picone, and Luca Veltri.
© 2019 John Wiley & Sons Ltd. Published 2019 by John Wiley & Sons Ltd.

8 Bormann, C., Lemay, S., Tschofenig, H., Hartke, K., Silverajan, B., and Raymor, B. (2017) CoAP (Constrained Application Protocol) over TCP, TLS, and WebSockets, Internet draft draft-ietf-core-coap-tcp-tls-09, Internet Engineering Task Force. URL `https://datatracker.ietf.org/doc/html/draft-ietf-core-coap-tcp-tls-09`, work in progress.

9 Hartke, K. (2015), Observing Resources in the Constrained Application Protocol (CoAP), RFC 7641, doi:10.17487/RFC7641. URL `https://rfc-editor.org/rfc/rfc7641.txt`.

10 Bormann, C. and Shelby, Z. (2016), Block-Wise Transfers in the Constrained Application Protocol (CoAP), RFC 7959. URL `http://www.ietf.org/rfc/rfc7959.txt`.

11 Rahman, A. and Dijk, E. (2014), Group Communication for the Constrained Application Protocol (CoAP), RFC 7390 (Experimental). URL `http://www.ietf.org/rfc/rfc7390.txt`.

12 Nottingham, M. (2010), Web Linking, RFC 5988. URL `http://www.ietf.org/rfc/rfc5988.txt`.

13 Shelby, Z., Constrained RESTful Environments (CoRE) Link Format, RFC 6690.

14 Shelby, Z., Koster, M., Bormann, C., van der Stok, P., and Amsuess, C. (2018) Core resource directory, Internet-Draft draft-ietf-core-resource-directory-13, IETF Secretariat. URL `http://www.ietf.org/internet-drafts/draft-ietf-core-resource-directory-13.txt`.

15 Castellani, A., Loreto, S., Rahman, A., Fossati, T., and Dijk, E. (2017), Guidelines for Mapping Implementations: HTTP to the Constrained Application Protocol (CoAP), RFC 8075. URL `http://www.ietf.org/rfc/rfc8075.txt`.

16 Schulzrinne, H., Casner, S., Frederick, R., and Jacobson, V. (2003), RTP: A Transport Protocol for Real-Time Applications, RFC 3550. URL `http://www.ietf.org/rfc/rfc3550.txt`, updated by RFCs 5506, 5761, 6051, 6222, 7022, 7160, 7164, 8083, 8108.

17 Handley, M., Jacobson, V., and Perkins, C. (2006), SDP: Session Description Protocol, RFC 4566 (proposed standard). URL `http://www.ietf.org/rfc/rfc4566.txt`.

18 Roach, A.B. (2002), Session Initiation Protocol (SIP)-Specific Event Notification, RFC 3265. URL `http://www.ietf.org/rfc/rfc3265.txt`, obsoleted by RFC 6665, updated by RFCs 5367, 5727, 6446.

19 IETF IPv6 over Low Power WPAN. URL `http://tools.ietf.org/wg/6lowpan/`.

20 IETF Routing Over Low Power and Lossy Networks Working Group. URL `http://tools.ietf.org/wg/roll/`.

21 IETF Constrained RESTful Environments Working Group. URL `http://tools.ietf.org/wg/core/`.

22 mjSIP Project. URL `http://mjsip.org`.

23 Cirani, S., Picone, M., and Veltri, L. (2014) mjCoAP: An open-source lightweight Java CoAP library for Internet of Things applications, in *Workshop on Interoperability and Open-Source Solutions for the Internet of Things, in conjunction with SoftCOM 2014: The 22nd International Conference on Software, Telecommunications and Computer Networks*, Split, Croatia.

24 mjCoAP, URL `http://netsec.unipr.it/project/mjcoap`.

25 Cirani, S., Picone, M., and Veltri, L. (2013) Cosip: A constrained session initiation protocol for the internet of things, in *Advances in Service-Oriented and Cloud Computing: Workshops of ESOCC 2013, Málaga, Spain, September 11–13, 2013, Revised Selected Papers*, Springer, Berlin, Heidelberg, pp. 13–24.

26 Cirani, S., Picone, M., and Veltri, L. (2013) A session initiation protocol for the Internet of Things. *Scalable Computing: Practice and Experience*, **14** (4), 249–263.

27 CoSIP Project. URL `http://netsec.unipr.it/project/cosip`.

28 Berners-Lee, T., Fielding, R., and Masinter, L. (2005), Uniform Resource Identifier (URI): Generic Syntax, RFC 3986. URL `http://www.ietf.org/rfc/rfc3986.txt`, updated by RFCs 6874, 7320.

29 Fielding, R.T. (2000) *Architectural Styles and the Design of Network-based Software Architectures*, Phd thesis, University of California. URL `http://www.ics.uci.edu/~fielding/pubs/dissertation/top.htm`.

30 Mitra, N. and Lafon, Y. (2007), SOAP Version 1.2 Part 0: Primer (2nd edn), World Wide Web Consortium, Recommendation REC-soap12-part0-20070427.

31 Hadley, M. (2009), Web Application Description Language, World Wide Web Consortium.

32 Kovatsch, M., Duquennoy, S., and Dunkels, A. (2011) A low-power CoAP for Contiki, in *IEEE 8th International Conference on Mobile Adhoc and Sensor Systems (MASS), 2011.*

33 mjSIP project (2013), mjUA: mjSIP User Agent (UA). URL `http://www.mjcoap.org/ua.`

34 Veltri, L. (2014), mjCoAP extension for session initiation. URL `http://www.mjcoap.org/projects/session.`

35 Cirani, S., Picone, M., and Veltri, L. (2014) A session initiation protocol for the Internet of Things. *Scalable Computing: Practice and Experience,* **14** (4), 249–263.

36 Hartke, K. (2015) Observing resources in the constrained application protocol (CoAP), RFC 7641, Internet Engineering Task Force.

37 Gallart, V., Felici-Castell, S., Delamo, M., Foster, A., and Perez, J. (2011) Evaluation of a real, low cost, urban WSN deployment for accurate environmental monitoring, in *IEEE 8th International Conference on Mobile Adhoc and Sensor Systems* (MASS).

38 Ruichao, L., Jing, H., and Lianfeng, S. (2009) Design and implementation of a video surveillance system based on 3G network, in *International Conference on Wireless Communications Signal Processing* (WCSP 2009).

39 Winter, T., Thubert, P., Brandt, A., Hui, J., Kelsey, R., Levis, P., Pister, K., Struik, R., Vasseur, J., and Alexander, R. (2012), RPL: IPv6 Routing Protocol for Low-Power and Lossy Networks, RFC 6550. URL `http://www.ietf.org/rfc/rfc6550.txt.`

40 Dunkels, A., Gronvall, B., and Voigt, T. (2004) Contiki – a lightweight and flexible operating system for tiny networked sensors, in *29th Annual IEEE International Conference on Local Computer Networks,* Tampa, FL, USA.

41 IEEE (2003) Local and Metropolitan Area Networks – Specific requirements– Part 11: Wireless LAN Medium Access Control (MAC) and Physical Layer (PHY) Specifications Amendment 4: Further Higher Data Rate Extension in the 2.4 GHz Band, 802.11g-2003, IEEE.

42 IEEE (2009) Local and Metropolitan Area Networks – Specific Requirements– Part 11: Wireless LAN Medium Access Control (MAC) and Physical Layer (PHY) Specifications Amendment 5: Enhancements for Higher Throughput, 802.11n-2009, IEEE.

43 IEEE (2011) Local and Metropolitan Area Networks – Specific requirements – Part 11: Wireless LAN Medium Access Control

(MAC) and Physical Layer (PHY) Specifications Amendment 10: Mesh Networking, 802.11s-2011, IEEE.

44 Gummeson, J., Ganesan, D., Corner, M., and Shenoy, P. (2010) An adaptive link layer for heterogeneous multi-radio mobile sensor networks. *IEEE J. Select. Areas Commun.*, **28** (7), 1094–1104.

45 Sengul, C., Bakht, M., Harris, A.F., Abdelzaher, T., and Kravets, R. (2008) Improving energy conservation using bulk transmission over high-power radios in sensor networks, in *Proceedings of the 28th International Conference on Distributed Computing Systems* (ICDCS 08), Beijing, China.

46 Wan, C.Y., Eisenman, S.B., Campbell, A.T., and Crowcroft, J. (2005) Siphon: Overload traffic management using multi-radio virtual sinks in sensor networks, in *Proceedings of the 3rd International Conference on Embedded Networked Sensor Systems* (SenSys 05), San Diego, CA, USA.

47 Stathopoulos, T., Lukac, M., McIntire, D., Heidemann, J., Estrin, D., and Kaiser, W. (2007) End-to-end routing for dual-radio sensor networks, in *26th IEEE International Conference on Computer Communications* (INFOCOM 2007), Anchorage, AK, USA.

48 Jurdak, R., Klues, K., Kusy, B., Richter, C., Langendoen, K., and Brunig, M. (2011) Opal: A multiradio platform for high throughput wireless sensor networks. *IEEE Embedded Systems Letters*, **3** (4), 121–124.

49 Dunkels, A. (2011) The ContikiMAC Radio Duty Cycling Protocol, Tech. Rep. T2011:13, Swedish Institute of Computer Science. URL http://dunkels.com/adam/dunkels11contikimac.pdf.

50 Mockapetris, P. (1987), Domain names – concepts and facilities, RFC 1034. URL http://www.ietf.org/rfc/rfc1034.txt, updated by RFCs 1101, 1183, 1348, 1876, 1982, 2065, 2181, 2308, 2535, 4033, 4034, 4035, 4343, 4035, 4592, 5936, 8020.

51 Mockapetris, P. (1987), Domain names - implementation and specification, RFC 1035. URL http://www.ietf.org/rfc/rfc1035.txt, updated by RFCs 1101, 1183, 1348, 1876, 1982, 1995, 1996, 2065, 2136, 2181, 2137, 2308, 2535, 2673, 2845, 3425, 3658, 4033, 4034, 4035, 4343, 5936, 5966, 6604, 7766.

52 Freed, N. and Borenstein, N. (1996), Multipurpose Internet Mail Extensions (MIME) Part One: Format of Internet Message Bodies, RFC 2045. URL http://www.ietf.org/rfc/rfc2045.txt, updated by RFCs 2184, 2231, 5335, 6532.

53 Nottingham, M. and Sayre, R. (2005), The Atom Syndication Format, RFC 4287. URL http://www.ietf.org/rfc/rfc4287.txt, updated by RFC 5988.

54 Crocker, D. and Overell, P. (2008), Augmented BNF for Syntax Specifications: ABNF, RFC 5234. URL http://www.ietf.org/rfc/rfc5234.txt, updated by RFC 7405.

55 Shelby, Z., Vial, M., Koster, M., Groves, C., Zhu, J., and Silverajan, B. (2018) Reusable Interface Definitions for Constrained RESTful Environments, Internet-Draft draft-ietf-core-interfaces-11, IETF Secretariat. URL http://www.ietf.org/internet-drafts/draft-ietf-core-interfaces-11.txt.

56 Jennings, C., Shelby, Z., Arkko, J., Keranen, A., and Bormann, C. (2018) Media Types for Sensor Measurement Lists (SenML), Internet-Draft draft-ietf-core-senml-14, IETF Secretariat. URL http://www.ietf.org/internet-drafts/draft-ietf-core-senml-14.txt.

57 UPnP forums. URL http://www.upnp.org/.

58 Guttman, E., Perkins, C., and Veizades, J. (1999) Service Location Protocol, Version 2, RFC 2608, IETF. URL http://tools.ietf.org/html/rfc2608.

59 Guttman, E. (2002) Vendor Extensions for Service Location Protocol, Version 2, RFC 3224, IETF. URL http://tools.ietf.org/html/rfc3224.

60 Zeroconf website. URL http://www.zeroconf.org/.

61 Cheshire, S. and Krochmal, M. (2013) Multicast DNS, RFC 6762, IETF. URL http://tools.ietf.org/html/rfc6762.

62 Cheshire, S. and Krochmal, M. (2013) DNS-Based Service Discovery, RFC 6763, IETF. URL http://tools.ietf.org/html/rfc6763.

63 Piax website. URL http://www.piax.org/en.

64 Kaneko, Y., Harumoto, K., Fukumura, S., Shimojo, S., and Nishio, S. (2005) A location-based peer-to-peer network for context-aware services in a ubiquitous environment, in *Applications and the Internet Workshops, 2005. Saint Workshops 2005. The 2005 Symposium on*.

65 Busnel, Y., Bertier, M., and Kermarrec, A.M. (2008) Solist or how to look for a needle in a haystack? A lightweight multi-overlay structure for wireless sensor networks, in *IEEE WiMob '08*, Avignon, France.

66 Gutierrez, G., Mejias, B., Van Roy, P., Velasco, D., and Torres, J. (2008) WSN and P2P: A self-managing marriage, in *2nd IEEE International Conference on Self-Adaptive and Self-Organizing Systems Workshops* (SASOW 2008).

67 Leguay, J., Lopez-Ramos, M., Jean-Marie, K., and Conan, V. (2008) An efficient service oriented architecture for heterogeneous and dynamic wireless sensor networks, in *33rd IEEE Conference on Local Computer Networks, 2008.* (LCN 2008).

68 Kovacevic, A., Ansari, J., and Mahonen, P. (2010) NanoSD: A flexible service discovery protocol for dynamic and heterogeneous wireless sensor networks, in *Mobile Ad-hoc and Sensor Networks (MSN), 2010 Sixth International Conference on.*

69 Mayer, S. and Guinard, D. (2011) An extensible discovery service for smart things, in *Proceedings of the Second International Workshop on Web of Things* (WOT '11).

70 Butt, T.A., Phillips, I., Guan, L., and Oikonomou, G. (2012) TRENDY: an adaptive and context-aware service discovery protocol for 6LoWPANs, in *Proceedings of the Third International Workshop on the Web of Things* (WOT '12), ACM, New York, NY, USA.

71 Jara, A., Lopez, P., Fernandez, D., Castillo, J., Zamora, M., and Skarmeta, A. (2013) Mobile Digcovery: A global service discovery for the internet of things, in *27th International Conference on Advanced Information Networking and Applications Workshops* (WAINA), 2013.

72 Paganelli, F. and Parlanti, D. (2012) A DHT-based discovery service for the Internet of Things. *J. Comp. Netw. Communic.*, **2012**. URL https://www.hindawi.com/journals/jcnc/2012/107041/cta/.

73 Schoenemann, N., Fischbach, K., and Schoder, D. (2009) P2P architecture for ubiquitous supply chain systems, in *ECIS* (eds S. Newell, E.A. Whitley, N. Pouloudi, J. Wareham, and L. Mathiassen).

74 Shrestha, S., Kim, D.S., Lee, S., and Park, J.S. (2010) A peer-to-peer RFID resolution framework for supply chain network, in *Future Networks, 2010. ICFN '10. Second International Conference on.*

75 Manzanares-Lopez, P., Muñoz-Gea, J.P., Malgosa-Sanahuja, J., and Sanchez-Aarnoutse, J.C. (2011) An efficient distributed discovery

service for EPCglobal network in nested package scenarios. *J. Netw. Comput. Appl.*, **34** (3), 925–937.

76 Rahman, A. and Dijk, E. (2014) Group Communication for the Constrained Application Protocol (CoAP), RFC 7390, RFC Editor. URL http://www.rfc-editor.org/rfc/rfc7390.txt.

77 Stoica, I., Morris, R., Karger, D., Kaashoek, M.F., and Balakrishnan, H. (2001) Chord: A scalable peer-to-peer lookup service for internet applications. *SIGCOMM Comput. Commun. Rev.*, **31** (4), 149–160. URL http://doi.acm.org/10.1145/964723.383071.

78 Maymounkov, P. and Mazières, D. (2002) Kademlia: A peer-to-peer information system based on the XOR metric, in *Proceedings of the 1st International Workshop on Peer-to-Peer Systems*, Springer-Verlag, London. URL http://dl.acm.org/citation.cfm?id=646334.687801.

79 Yulin, N., Huayou, S., Weiping, L., and Zhong, C. (2010) PDUS: P2P-based distributed UDDI service discovery approach, in *International Conference on Service Sciences (ICSS), 2010.*

80 Kaffille, S., Loesing, K., and Wirtz, G. (2005) Distributed service discovery with guarantees in peer-to-peer networks using distributed hashtables, in *PDPTA*, CSREA Press.

81 Cheshire, S., Aboba, B., and Guttman, E. (2005), Dynamic Configuration of IPv4 Link-Local Addresses, RFC 3927. URL http://www.ietf.org/rfc/rfc3927.txt.

82 Cheshire, S. and Krochmal, M. (2013), Multicast DNS, RFC 6762. URL http://www.ietf.org/rfc/rfc6762.txt.

83 Cheshire, S. and Krochmal, M. (2013), DNS-Based Service Discovery, RFC 6763. URL http://www.ietf.org/rfc/rfc6763.txt.

84 Cirani, S. and Veltri, L. (2008) Implementation of a framework for a DHT-based distributed location service, in *16th International Conference on Software, Telecommunications and Computer Networks, 2008.* (SoftCOM 2008).

85 Picone, M., Amoretti, M., and Zanichelli, F. (2010) GeoKad: A P2P distributed localization protocol, in *Pervasive Computing and Communications Workshops (PERCOM Workshops), 2010 8th IEEE International Conference on.*

86 Picone, M., Amoretti, M., and Zanichelli, F. (2011) Proactive neighbor localization based on distributed geographic table. *Int. J Pervasive Comput. Commun.*, **7**, 240–263.

87 Bryan, D., Lowekamp, B., and Jennings, C. (2007) dSIP: A P2P Approach to SIP Registration and Resource Location, Internet draft draft-bryan-p2psip-dsip-00, IETF. URL http://tools .ietf.org/id/draft-bryan-p2psip-dsip-00.txt.

88 Jennings, C., Lowekamp, B., Rescorla, E., Baset, S., and Schulzrinne, H. (2014) REsource LOcation And Discovery (RELOAD) Base Protocol, RFC 6940, RFC Editor.

89 Gonizzi, P., Ferrari, G., Gay, V., and Leguay, J. (2013) Data dissemination scheme for distributed storage for IoT observation systems at large scale. *Inf. Fusion*, **22**, 16–25.

90 Droms, R. (1997) Dynamic Host Configuration Protocol, RFC 2131. URL http://www.ietf.org/rfc/rfc2131.txt, updated by RFCs 3396, 4361, 5494, 6842.

91 Picone, M., Amoretti, M., and Zanichelli, F. (2011) Evaluating the robustness of the DGT approach for smartphone-based vehicular networks, in *Proceedings of IEEE 36th Conference on Local Computer Networks* (LCN 2011), IEEE, Bonn, Germany.

92 Picone, M., Amoretti, M., and Zanichelli, F. (2012) A decentralized smartphone based traffic information system, in *Proceedings of the 2012 IEEE Intelligent Vehicles Symposium*, IEEE, Alcalá de Henares, Spain.

93 Cirani, S. and Veltri, L. (2007), A Kademlia-based DHT for Resource Lookup in P2PSIP, Obsolete Internet draft.

94 Kovatsch, M., Lanter, M., and Shelby, Z. (2014) Californium: scalable cloud services for the internet of things with CoAP, in *Proceedings of the 4th International Conference on the Internet of Things* (IoT 2014), Cambridge, MA, USA.

95 Weisstein, E.W., Least squares fitting – logarithmic. URL http://mathworld.wolfram.com/ LeastSquaresFittingLogarithmic.html.

96 Hui, J. and Kelsey, R. (2016) Multicast protocol for low-power and lossy networks (mpl), Internet Requests for Comments RFC 7731, RFC Editor.

97 Levis, P., Patel, N., Culler, D., and Shenker, S. (2004) Trickle: A self-regulating algorithm for code propagation and maintenance in wireless sensor networks, in *Proceedings of the 1st Conference on Symposium on Networked Systems Design and Implementation*

(NSDI04), vol. **1**, USENIX Association, Berkeley, CA, USA, vol. 1. URL `http://dl.acm.org/citation.cfm?id=1251175 .1251177`.

98 Jung, M. and Kastner, W. (2013) Efficient group communication based on web services for reliable control in wireless automation, in *39th Annual Conference of the IEEE Industrial Electronics Society, IECON 2013*.

99 Oikonomou, G. and Phillips, I. (2012) Stateless multicast forwarding with RPL in 6LoWPAN sensor networks, in *IEEE International Conference on Pervasive Computing and Communications Workshops (PERCOM Workshops), 2012*.

100 Bloom, B.H. (1970) Space/time trade-offs in hash coding with allowable errors. *Commun. ACM*, **13** (7), 422–426. URL `http://doi.acm.org/10.1145/362686.362692`.

101 Bender, M.A., Farach-Colton, M., Johnson, R., Kuszmaul, B.C., Medjedovic, D., Montes, P., Shetty, P., Spillane, R.P., and Zadok, E. (2011) Don't thrash: How to cache your hash on flash, in *Proceedings of the 3rd USENIX Conference on Hot Topics in Storage and File Systems* (HotStorage '11), USENIX Association, Berkeley, CA, USA. URL `http://dl.acm.org/citation.cfm? id=2002218.2002219`.

102 Heer, T., Garcia-Morchon, O., Hummen, R., Keoh, S.L., Kumar, S.S., and Wehrle, K. (2011) Security challenges in the IP-based Internet of Things. *Wirel. Pers. Commun.*, **61** (3), 527–542. URL `http://dx.doi.org/10.1007/s11277-011-0385-5`.

103 Garcia-Morchon, O., Keoh, S., Kumar, S., Hummen, R., and Struik, R. (2012) Security Considerations in the IP-based Internet of Things, Internet draft draft-garcia-core-security-04, IETF. URL `http://tools.ietf.org/id/draft-garcia-core-security-04`.

104 Bormann, C. (2012) Guidance for Light-Weight Implementations of the Internet Protocol Suite, Internet draft draft-ietf-lwig-guidance-02, IETF. URL `http://tools.ietf .org/html/draft-ietf-lwig-guidance`.

105 IETF Light-Weight Implementation Guidance. URL `http:// tools.ietf.org/wg/lwig/`.

106 Kent, S. and Atkinson, R. (1998), Security Architecture for the Internet Protocol, RFC 2401. URL `http://www.ietf.org/`

`rfc/rfc2401.txt`, obsoleted by RFC 4301, updated by RFC 3168.

107 Raza, S., Duquennoy, S., Chung, T., Yazar, D., Voigt, T., and Roedig, U. (2011) Securing communication in 6LoWPAN with compressed IPsec, in *Proceedings of the International Conference on Distributed Computing in Sensor Systems* (IEEE DCOSS 2011). URL `http://www.simonduquennoy.net/papers/raza11securing.pdf`.

108 Moskowitz, R., Heer, T., Jokela, P., and Henderson, T. (2015) Host Identity Protocol Version 2 (HIPv2), RFC 7401, RFC Editor. URL `http://www.rfc-editor.org/rfc/rfc7401.txt`.

109 Jokela, P., Moskowitz, R., and Melen, J. (2015) Using the Encapsulating Security Payload (ESP) Transport Format with the Host Identity Protocol (HIP), Tech. Rep., IETF.

110 Dierks, T. and Rescorla, E. (2008), The Transport Layer Security (TLS) Protocol Version 1.2, RFC 5246. URL `http://www.ietf.org/rfc/rfc5246.txt`, updated by RFCs 5746, 5878, 6176, 7465, 7507, 7568, 7627, 7685, 7905, 7919.

111 Rescorla, E. and Modadugu, N. (2012), Datagram Transport Layer Security Version 1.2, RFC 6347. URL `http://www.ietf.org/rfc/rfc6347.txt`, updated by RFCs 7507, 7905.

112 Raza, S., Trabalza, D., and Voigt, T. (2012) 6LoWPAN Compressed DTLS for CoAP, in *Proceedings of the 8th IEEE International Conference on Distributed Computing in Sensor Systems* (IEEE DCOSS 2011), Hangzhou, China.

113 Brachmann, M., Keoh, S., Morchon, O., and Kumar, S. (2012) End-to-end transport security in the IP-Based Internet of Things, in *21st International Conference on Computer Communications and Networks* (ICCCN), *2012*.

114 Brachmann, M., Morchon, O., Keoh, S., and Kumar, S. (2012) Security considerations around end-to-end security in the IP-based Internet of Things, in *Workshop on Smart Object Security, in conjunction with IETF83*.

115 Ramsdell, B. and Turner, S. (2010), Secure/Multipurpose Internet Mail Extensions (S/MIME) Version 3.2 Message Specification, RFC 5751. URL `http://www.ietf.org/rfc/rfc5751.txt`.

116 Baugher, M., McGrew, D., Naslund, M., Carrara, E., and Norrman, K. (2004), The Secure Real-time Transport Protocol (SRTP), RFC 3711. URL `http://www.ietf.org/rfc/rfc3711.txt`, updated by RFCs 5506, 6904.

117 Daemen, J. and Rijmen, V. (2002) *The Design of Rijndael*, Springer-Verlag New York, Inc., Secaucus, NJ, USA.

118 Eisenbarth, T., Kumar, S., Paar, C., Poschmann, A., and Uhsadel, L. (2007) A survey of lightweight-cryptography implementations. *IEEE Des. Test*, **24** (6), 522–533.

119 Rinne, S., Eisenbarth, T., and Paar, C. (2007) Performance analysis of contemporary light-weight block ciphers on 8-bit microcontrollers, in *ECRYPT Workshop SPEED – Software Performance Enhancement for Encryption and Decryption*.

120 Wheeler, D. and Needham, R. (1995) TEA, a tiny encryption algorithm, in *Fast Software Encryption, Lecture Notes in Computer Science*, vol. **1008** (ed. B. Preneel), Springer, pp. 363–366.

121 Needham, R.M. and Wheeler, D.J. (1997) TEA Extensions, Tech. Rep., University of Cambridge.

122 Standaert, F.X., Piret, G., Gershenfeld, N., and Quisquater, J.J. (2006) SEA: A scalable encryption algorithm for small embedded applications, in *Smart Card Research and Advanced Applications, Lecture Notes in Computer Science*, vol. 3928 (eds J. Domingo-Ferrer, J. Posegga, and D. Schreckling), Springer, pp. 222–236.

123 Bogdanov, A., Knudsen, L., Leander, G., Paar, C., Poschmann, A., Robshaw, M., Seurin, Y., and Vikkelsoe, C. (2007) PRESENT: An ultra-lightweight block cipher, in *Cryptographic Hardware and Embedded Systems* (CHES 2007), *Lecture Notes in Computer Science*, vol. **4727** (eds P. Paillier and I. Verbauwhede), Springer, pp. 450–466.

124 ISO/IEC (2012) Information Technology – Security Techniques – Lightweight Cryptography – Part 2: Block Ciphers, ISO/IEC standard 29192-2:2012, ISO, Geneva, Switzerland.

125 Hong, D., Sung, J., Hong, S., Lim, J., Lee, S., Koo, B., Lee, C., Chang, D., Lee, J., Jeong, K., Kim, H., Kim, J., and Chee, S. (2006) HIGHT: A new block cipher suitable for low-resource device, in *Cryptographic Hardware and Embedded Systems – CHES 2006, 8th International Workshop, Lecture Notes in Computer Science*, vol. **4249**, Springer, *Lecture Notes in Computer Science*, vol. **4249**. URL http://www.iacr.org/cryptodb/archive/2006/CHES/04/04.pdf.

126 Atmel AVR 8-bit microcontrollers. URL http://www.atmel.it/products/microcontrollers/avr/default.aspx.

127 Feldhofer, M., Dominikus, S., and Wolkerstorfer, J. (2004) Strong authentication for RFID systems using the AES algorithm, in *Cryptographic Hardware and Embedded Systems – CHES 2004, Lecture Notes in Computer Science*, vol. **3156** (eds M. Joye and J.J. Quisquater), Springer, pp. 357–370.

128 Feldhofer, M., Wolkerstorfer, J., and Rijmen, V. (2005) AES implementation on a grain of sand. *IEEE Proc. Info. Security,*, **152** (1), 13–20.

129 Kaps, J.P. (2008) Chai-tea, cryptographic hardware implementations of xTEA, in *Progress in Cryptology* (INDOCRYPT 2008), *Lecture Notes in Computer Science*, vol. **5365** (eds D. Chowdhury, V. Rijmen, and A. Das), Springer, pp. 363–375.

130 Macé, F., St, F.X., and Quisquater, J.J. (2007), ASIC implementations of the block cipher SEA for constrained applications. URL http://citeseerx.ist.psu.edu/viewdoc/summary?doi=10.1.1.88.926.

131 Plos, T., Dobraunig, C., Hofinger, M., Oprisnik, A., Wiesmeier, C., and Wiesmeier, J. (2012) Compact hardware implementations of the block ciphers mCrypton, NOEKEON, and SEA, in *Progress in Cryptology* (INDOCRYPT 2012), *Lecture Notes in Computer Science*, vol. **7668** (eds S. Galbraith and M. Nandi), Springer, pp. 358–377.

132 Rivest, R., Shamir, A., and Adleman, L. (1978) A method for obtaining digital signatures and public-key cryptosystems. *Commun. ACM*, **21**, 120–126.

133 Koblitz, N. (1987) Elliptic curve cryptosystems. *Math. Comput.*, **48** (177), 203–209.

134 American National Standards Institute. URL http://www.ansi.org.

135 Institute of Electrical and Electronics Engineers. URL http://www.ieee.org.

136 International Organization for Standardization. URL http://www.ieee.org.

137 Standards for Efficient Cryptography Group. URL http://secs.org.

138 National Institute of Standards and Technology. URL http://www.nist.gov.

139 Sethi, M., Arkko, A., Keranen, A., and Rissanen, H. (2012) Practical Considerations and Implementation Experiences in Securing Smart Object Networks, Internet draft

draft-aks-crypto-sensors-02, IETF. URL `http://tools.ietf`
`.org/html/draft-aks-crypto-sensors-02.`

140 Rivest, R. (1992) RFC 1321: The MD5 message-digest algorithm. *The Internet Engineering Task Force (IETF).*

141 Eastlake, D.E. and Jones, P.E., US Secure Hash Algorithm 1 (SHA1), `http://www.ietf.org/rfc/rfc3174.txt.`

142 Bogdanov, A., Leander, G., Paar, C., Poschmann, A., Robshaw, M.J., and Seurin, Y. (2008) Hash functions and RFID tags: Mind the gap, in *Proceedings of the 10th International Workshop on Cryptographic Hardware and Embedded Systems* (CHES '08), Springer-Verlag, Berlin, Heidelberg.

143 Guo, J., Peyrin, T., and Poschmann, A. (2011) The photon family of lightweight hash functions, in *Proceedings of the 31st Annual Conference on Advances in Cryptology* (CRYPTO'11), Springer-Verlag, Berlin, Heidelberg. URL `http://dl.acm`
`.org/citation.cfm?id=2033036.2033053.`

144 Bertoni, G., Daemen, J., Peeters, M., Assche, G.V., Bertoni, G., Daemen, J., Peeters, M., and Assche, G.V. (2007), Sponge functions. URL `https://keccak.team/files/CSF-0.1.pdf.`

145 Bogdanov, A., Knežević, M., Leander, G., Toz, D., Varici, K., and Verbauwhede, I. (2011) Spongent: a lightweight hash function, in *Proceedings of the 13th International Conference on Cryptographic Hardware and Embedded Systems* (CHES'11), Springer-Verlag, Berlin, Heidelberg. URL `http://dl.acm.org/citation`
`.cfm?id=2044928.2044957.`

146 Aumasson, J.P., Henzen, L., Meier, W., and Naya-Plasencia, M. (2010) Quark: a lightweight hash, in *Proceedings of the 12th International Conference on Cryptographic Hardware and Embedded Systems* (CHES'10), Springer-Verlag, Berlin, Heidelberg. URL `http://dl.acm.org/citation.cfm?id=1881511`
`.1881513.`

147 Bertoni, G., Daemen, J., Peeters, M., and Assche, G.V. (2009), Keccak specifications. URL `http://keccak.noekeon.org/`
`Keccak-specifications.pdf.`

148 NIST, Sha3 competition. URL `http://csrc.nist.gov/`
`groups/ST/hash/sha-3.`

149 Shamir, A. (2008) SQUASH – A new MAC with provable security properties for highly constrained devices such as RFID tags, in *Fast Software Encryption* (ed. K. Nyberg), Springer-Verlag, Berlin, Heidelberg, pp. 144–157.

150 El Gamal, T. (1985) A public key cryptosystem and a signature scheme based on discrete logarithms, in *Proceedings of CRYPTO 84 on Advances in cryptology*, Springer-Verlag New York, Inc., New York, NY, USA. URL `http://dl.acm.org/citation.cfm?id=19478.19480`.

151 Paillier, P. (1999) Public-key cryptosystems based on composite degree residuosity classes, in *Advances in Cryptology* (EUROCRYPT 99), *Lecture Notes in Computer Science*, vol. 1592 (ed. J. Stern), Springer, pp. 223–238.

152 Diffie, W. and Hellman, M. (2006), New directions in cryptography.

153 Harkins, D. and Carrel, D. (1998), The Internet Key Exchange (IKE), RFC 2409. URL `http://www.ietf.org/rfc/rfc2409.txt`, obsoleted by RFC 4306, updated by RFC 4109.

154 Blundo, C., Santis, A.D., Herzberg, A., Kutten, S., Vaccaro, U., and Yung, M. (1993) Perfectly-secure key distribution for dynamic conferences, in *Proceedings of the 12th Annual International Cryptology Conference on Advances in Cryptology* (CRYPTO '92), Springer-Verlag, London, UK, UK, CRYPTO '92. URL `http://dl.acm.org/citation.cfm?id=646757.705531`.

155 Liu, D. and Ning, P. Establishing pairwise keys in distributed sensor networks, in *Proceedings of the 10th ACM Conference on Computer and Communications Security* (CCS '03).

156 Perrig, A., Szewczyk, R., Tygar, J.D., Wen, V., and Culler, D.E. (2002) Spins: security protocols for sensor networks. *Wirel. Netw.*, **8** (5), 521–534.

157 Chan, H., Perrig, A., and Song, D. Random key predistribution schemes for sensor networks, in *Proceedings of the 2003 IEEE Symposium on Security and Privacy* (SP '03).

158 Wong, C.K., Gouda, M., and Lam, S. (2000) Secure group communications using key graphs. *IEEE/ACM Trans. Netw.*, **8** (1), 16–30.

159 Micciancio, D. and Panjwani, S. (2008) Optimal communication complexity of generic multicast key distribution. *IEEE/ACM Trans. Netw.*, **16**.

160 Keoh, S., Garcia-Morchon, O., Kumar, S., and Dijk, S. (2012) DTLS-based Multicast Security for Low-Power and Lossy Networks (LLNs), Internet draft draft-keoh-tls-multicast-security-00, IETF. URL `http://tools.ietf.org/id/draft-keoh-tls-multicast-security-00`.

161 Berkovits, S. (1991) How to broadcast a secret, in *Proc. of the Intl. Conference on Theory and application of cryptographic techniques* (EUROCRYPT), Springer-Verlag, Brighton, UK.

162 Naor, D., Naor, M., and Lotspiech, J. (2001) Revocation and tracing schemes for stateless receivers, in *Advances in Cryptology* (CRYPTO), *Lecture Notes in Computer Science*, vol. 2139 (ed. J. Kilian), Springer Berlin / Heidelberg, pp. 41–62.

163 Ballardie, A. (1996), Scalable Multicast Key Distribution, RFC 1949 (Experimental). URL http://www.ietf.org/rfc/rfc1949.txt.

164 Lin, J., Huang, K., Lai, F., and Lee, H. (2009) Secure and efficient group key management with shared key derivation. *Comput. Standards Interf.*, **31** (1), 192–208.

165 Lee, P., Lui, J., and Yau, D. (2006) Distributed collaborative key agreement and authentication protocols for dynamic peer groups. *IEEE/ACM Trans. Netw.*, **14** (2), 263–276.

166 Kim, Y., Perrig, A., and Tsudik, G. (2004) Tree-based group key agreement. *ACM Trans. Inf. Syst. Secur.*, **7**, 60–96.

167 Briscoe, B. (1999) MARKS: Zero side effect multicast key management using arbitrarily revealed key sequences, in *Networked Group Communication, Lecture Notes in Computer Science*, vol. **1736** (eds L. Rizzo and S. Fdida), Springer Berlin / Heidelberg, pp. 301–320.

168 Sarikaya, B., Ohba, Y., Moskowitz, R., Cao, Z., and Cragie, R. (2012) Security Bootstrapping Solution for Resource-Constrained Devices, Internet draft draft-sarikaya-core-sbootstrapping-05, IETF. URL http://tools.ietf.org/html/draft-sarikaya-core-sbootstrapping.

169 Jennings, C. (2012) Transitive Trust Enrollment for Constrained Devices, Internet draft draft-jennings-core-transitive-trust-enrollment-01, IETF. URL http://tools.ietf.org/id/draft-jennings-core-transitive-trust-enrollment-01.

170 Garcia-Morchon, O. and Wehrle, K. (2010) Modular context-aware access control for medical sensor networks, in *Proceedings of the 15th ACM Symposium on Access Control Models and Technologies* (SACMAT '10), ACM, New York, NY, USA. URL http://doi.acm.org/10.1145/1809842.1809864.

171 Hammer-Lahav, E. (2010) The OAuth 1.0 Protocol, RFC 5849, IETF. URL http://www.ietf.org/rfc/rfc5849.txt, obsoleted by RFC 6749.

172 Hardt, D. (2012), The OAuth 2.0 Authorization Framework, RFC 6749. URL `http://www.ietf.org/rfc/rfc6749.txt`.

173 Hammer-Lahav, E. (2010) The OAuth 1.0 Protocol, RFC 5849, IETF. URL `http://www.ietf.org/rfc/rfc5849.txt`, obsoleted by RFC 6749.

174 Dierks, T. and Rescorla, E. The Transport Layer Security (TLS) Protocol Version 1.2, RFC 5246, IETF.

175 Hardt, D. The OAuth 2.0 Authorization Framework, RFC 6749, IETF.

176 IPSO Alliance. URL `http://www.ipso-alliance.org/`.

177 Connect All IP-based Smart Objects (CALIPSO) – FP7 EU Project. URL `http://www.ict-calipso.eu/`.

178 Worldsensing. URL `http://www.worldsensing.com/`.

179 Cirani, S., Ferrari, G., and Veltri, L. (2013) Enforcing security mechanisms in the IP-Based Internet of Things: An algorithmic overview. *Algorithms*, **6** (2), 197–226. URL `http://www.mdpi.com/1999-4893/6/2/197`.

180 Ning, H., Liu, H., and Yang, L. (2013) Cyberentity security in the Internet of Things. *Computer*, **46** (4), 46–53.

181 Yao, X., Han, X., Du, X., and Zhou, X. (2013) A lightweight multicast authentication mechanism for small scale IoT applications. *IEEE Sensors J.*, **13** (10), 3693–3701.

182 Lai, C., Li, H., Liang, X., Lu, R., Zhang, K., and Shen, X. (2014) CPAL: A conditional privacy-preserving authentication with access linkability for roaming service. *IEEE Internet of Things J.*, **1** (1), 46–57.

183 Li, F. and Xiong, P. (2013) Practical secure communication for integrating wireless sensor networks into the Internet of Things. *IEEE Sensors J.*, **13** (10), 3677–3684.

184 Forsberg, D., Ohba, Y., Patil, B., Tschofenig, H., and Yegin, A. (2008), Protocol for Carrying Authentication for Network Access (PANA), RFC 5191. URL `http://www.ietf.org/rfc/rfc5191.txt`, updated by RFC 5872.

185 Aboba, B., Blunk, L., Vollbrecht, J., Carlson, J., and Levkowetz, H. (2004), Extensible Authentication Protocol (EAP), RFC 3748. URL `http://www.ietf.org/rfc/rfc3748.txt`, updated by RFCs 5247, 7057.

186 Moreno-Sanchez, P., Marin-Lopez, R., and Vidal-Meca, F. (2014) An open source implementation of the protocol for carrying authentication for network access: OpenPANA. *Network, IEEE*, **28** (2), 49–55.

187 United States Department of Defense (1985) Department of Defense Trusted Computer System Evaluation Criteria, Tech. Rep., United States Department of Defense. URL http://csrc.nist.gov/publications/history/dod85.pdf.

188 Ferraiolo, D. and Kuhn, R. (1992) Role-based access control, in *Proceedings of the 15th NIST-NCSC National Computer Security Conference*, Baltimore, MD, USA.

189 Ferraiolo, D.F., Sandhu, R., Gavrila, S., Kuhn, D.R., and Chandramouli, R. (2001) Proposed NIST standard for role-based access control. *ACM Trans. Inf. Syst. Secur.*, **4** (3), 224–274.

190 Sandhu, R., Coyne, E., Feinstein, H., and Youman, C. (1996) Role-based access control models. *Computer*, **29** (2), 38 –47.

191 Yuan, E. and Tong, J. (2005) Attributed based access control (ABAC) for Web services, in *Proceedings of the 2005 IEEE International Conference on Web Services, 2005 (ICWS 2005)*.

192 Schiffman, J., Zhang, X., and Gibbs, S. (2010) DAuth: Fine-grained authorization delegation for distributed web application consumers, in *Proceedings of the 2010 IEEE International Symposium on Policies for Distributed Systems and Networks (POLICY)*.

193 Gerdes, S., Bergmann, O., and Bormann, C. (2014) Delegated CoAP Authentication and Authorization Framework (DCAF), Internet draft draft-gerdes-ace-dcaf-authorize-00, IETF. URL http://tools.ietf.org/html/draft-gerdes-ace-dcaf-authorize-00.

194 OpenID Foundation (2007) OpenID Authentication 2.0 – Final, Tech. Rep., OpenID Foundation. URL http://openid.net/specs/openid-authentication-2_0.html.

195 Rescorla, E. and Modadugu, N., Datagram Transport Layer Security Version 1.2, RFC 6347.

196 Kent, S. and Atkinson, R., Security Architecture for the Internet Protocol, RFC 2401.

197 Stanoevska-Slabeva, K. and Wozniak, T. (2010) *Grid and Cloud Computing: A Business Perspective on Technology and Applications*, Springer.

198 Mell, P.M. and Grance, T. (2011) The NIST Definition of Cloud Computing, Tech. Rep. SP 800-145, National Institute of Standards & Technology.

199 Milojičić, D., Llorente, I.M., and Montero, R.S. (2011) OpenNebula: A cloud management tool. *IEEE Internet Computing*, **15** (2), 11–14.

200 McAfee, A. and Brynjolfsson, E., Big Data: The Management Revolution. URL https://hbr.org/2012/10/big-data-the-management-revolution.

201 Hohpe, G. and Woolf, B. (2003) *Enterprise Integration Patterns: Designing, Building, and Deploying Messaging Solutions*, Addison-Wesley Longman Publishing Co., Inc., Boston, MA, USA.

202 Isaacson, C. (2009) *Software Pipelines and SOA: Releasing the Power of Multi-Core Processing*, Addison-Wesley Professional.

203 Assunção, M.D., Calheiros, R.N., Bianchi, S., Netto, M.A., and Buyya, R. (2015) Big Data computing and clouds: Trends and future directions. *Journal of Parallel and Distributed Computing*, **79-80**, 3–15.

204 IETF, The Internet Engineering Task Force, URL http://www.ietf.org.

205 Bonomi, F., Milito, R., Zhu, J., and Addepalli, S. Fog computing and its role in the internet of things, in *Proceedings of the First Edition of the MCC Workshop on Mobile Cloud Computing* (MCC '12).

206 NGINX, The High-performance Web Server and Reverse Proxy, URL http://wiki.nginx.org/Main.

207 Fielding, R.T. and Kaiser, G. (1997) The Apache HTTP server project. *IEEE Internet Computing*, **1** (4), 88–90.

208 Vilajosana, I., Llosa, J., Martinez, B., Domingo-Prieto, M., Angles, A., and Vilajosana, X. (2013) Bootstrapping smart cities through a self-sustainable model based on big data flows. *IEEE Commun. Mag.*, **51** (6), 128–134.

209 Collina, M., Corazza, G.E., and Vanelli-Coralli, A. (2012) Introducing the QEST broker: Scaling the IoT by bridging MQTT and REST, in *2012 IEEE 23rd International Symposium on Personal, Indoor and Mobile Radio Communications – (PIMRC)*.

210 Lagutin, D., Visala, K., Zahemszky, A., Burbridge, T., and Marias, G.F. (2010) Roles and security in a publish/subscribe network architecture, in *2010 IEEE Symposium on Computers and Communications* (ISCC).

211 Wang, C., Carzaniga, A., Evans, D., and Wolf, A.L. (2002) Security issues and requirements for Internet-scale publish-subscribe

systems, in *Proceedings of the 35th Annual Hawaii International Conference on System Sciences, 2002* (HICSS).

212 Raiciu, C. and Rosenblum, D.S. (2006) Enabling confidentiality in content-based publish/subscribe infrastructures, in *Securecomm and Workshops, 2006*.

213 Fremantle, P., Aziz, B., Kopecký, J., and Scott, P. (2014) Federated identity and access management for the Internet of Things, in *2014 International Workshop on Secure Internet of Things* (SIoT).

214 Bacon, J., Evans, D., Eyers, D.M., Migliavacca, M., Pietzuch, P., and Shand, B. (2010) Enforcing end-to-end application security in the cloud, in *Proceedings of Middleware 2010: ACM/IFIP/USENIX 11th International Middleware Conference, Bangalore, India, November 29–December 3, 2010.*, Springer, Berlin, Heidelberg.

215 Yannuzzi, M., Milito, R., Serral-Gracià, R., Montero, D., and Nemirovsky, M. (2014) Key ingredients in an IoT recipe: Fog computing, cloud computing, and more fog computing, in *2014 IEEE 19th International Workshop on Computer Aided Modeling and Design of Communication Links and Networks (CAMAD)*.

216 Docker. URL https://www.docker.com/.

217 Ismail, B.I., Goortani, E.M., Karim, M.B.A., Tat, W.M., Setapa, S., Luke, J.Y., and Hoe, O.H. (2015) Evaluation of Docker as edge computing platform, in *2015 IEEE Conference on Open Systems* (ICOS).

218 Xu, Y., Mahendran, V., and Radhakrishnan, S. (2016) SDN Docker: Enabling application auto-docking/undocking in edge switch, in *2016 IEEE Conference on Computer Communications Workshops* (INFOCOM WKSHPS).

219 Ramalho, F. and Neto, A. (2016) Virtualization at the network edge: A performance comparison, in *2016 IEEE 17th International Symposium on A World of Wireless, Mobile and Multimedia Networks* (WoWMoM).

220 Morabito, R. (2016) A performance evaluation of container technologies on internet of things devices, in *2016 IEEE Conference on Computer Communications Workshops* (INFOCOM WKSHPS).

221 Aazam, M., Khan, I., Alsaffar, A., and Huh, E.N. (2014) Cloud of things: Integrating internet of things and cloud computing and the issues involved, in *2014 11th International Bhurban Conference on Applied Sciences and Technology* (IBCAST).

222 Aazam, M., Hung, P.P., and Huh, E.N. (2014) Smart gateway based communication for cloud of things, in *2014 IEEE Ninth International Conference on Intelligent Sensors, Sensor Networks and Information Processing* (ISSNIP).

223 Raspberry Pi Foundation. URL http://www.raspberrypi.org/.

224 Microsoft, Microsoft Azure – Cloud platform. URL http://azure.microsoft.com/it-it/.

225 Amazon Inc., Amazon EC2. URL http://aws.amazon.com/ec2/.

226 Rackspace, NASA, OpenStack Cloud Software – Open source software for building private and public clouds. URL https://www.openstack.org/.

227 Oracle Java VM, Java. URL https://www.oracle.com/java/index.html.

228 Dinakar Dhurjati, Sumant Kowshik, V.A. and Lattner, C. (2003) Memory safety without runtime checks or garbage collection, in *Proceedings of Languages Compilers and Tools for Embedded Systems 2003*, San Diego, CA. URL http://llvm.cs.uiuc.edu/pubs/2003-05-05-LCTES03-CodeSafety.html.

229 Bormann, C., Ersue, M., and Keranen, A. (2014), Terminology for Constrained-Node Networks, RFC 7228 (Informational). URL http://www.ietf.org/rfc/rfc7228.txt.

230 Mulligan, G. The 6LoWPAN architecture, in *Proceedings of the 4th Workshop on Embedded Networked Sensors* (EmNets '07).

231 Want, R., Schilit, B.N., and Jenson, S. (2015) Enabling the internet of things. *Computer*, **48** (1), 28–35.

232 Lea, R. (2014) Hypercat: An IoT Interoperability Specification, Tech. Rep., IoT Ecosystem Demonstrator Interoperability Working Group. URL http://eprints.lancs.ac.uk/id/eprint/69124.

233 Heggestuen, J. (2013) One in every 5 people in the world own a smartphone, one in every 17 own a tablet. *Business Insider*, **15 Dec.** URL http://www.businessinsider.com/smartphone-and-tablet-penetration-2013-10?IR=T.

234 Ham, T.H., It's a little freaky at first, but you get used to it. URL http://blog.ted.com/sergey-brin-with-google-glass-at-ted2013/.

235 Bojanova, I., Hurlburt, G., and Voas, J. (2014) Imagineering an internet of anything. *Computer*, **47** (6), 72–77.

236 Fettweis, G., Boche, H., Wiegand, T., Zielinski, E., Schotten, H., Merz, P., and Hirche, S. (2014) The Tactile Internet, Tech. Rep., ITU-T Technology Watch. URL `http://www.itu.int/dms_pub/itu-t/oth/23/01/T23010000230001PDFE.pdf`.

237 Heuer, J., Hund, J., and Pfaff, O. (2015) Toward the Web of Things: applying Web technologies to the physical world. *Computer*, **48** (5), 34–42.

238 Hardjono, T., Maler, E., Machulak, M., and Catalano, D. (2015) User-Managed Access (UMA) Profile of OAuth 2.0, Internet draft draft-hardjono-oauth-umacore-13, IETF. URL `https://tools.ietf.org/html/draft-hardjono-oauth-umacore-13.txt`.

Index

Internet of Things: Architectures, Protocols and Standards, First Edition.
Simone Cirani, Gianluigi Ferrari, Marco Picone, and Luca Veltri.
© 2019 John Wiley & Sons Ltd. Published 2019 by John Wiley & Sons Ltd.